T0344346

Materials and Innovative Product Development

Materials and Innovative Product Development

Using Common Sense

Gernot H. Gessinger

AMSTERDAM • BOSTON • HEIDELBERG • LONDON
NEW YORK • OXFORD • PARIS • SAN DIEGO
SAN FRANCISCO • SINGAPORE • SYDNEY • TOKYO

Butterworth-Heinemann is an imprint of Elsevier

Butterworth-Heinemann is an imprint of Elsevier
30 Corporate Drive, Suite 400, Burlington, MA 01803, USA
Linacre House, Jordan Hill, Oxford OX2 8DP, UK

Library of Congress Cataloging-in-Publication Data
Gessinger, G. H.
   Materials and innovative product development : from concept to market /
Gernot H. Gessinger. — 1st ed.
      p. cm.
   Includes bibliographical references and index.
   ISBN 978-1-85617-559-3 (hardcover)
1. Materials science. 2. Product management. 3. New products. 4. Research, Industrial–Management.
I. Title.
   TA403.6.G468 2009
   658.5′75—dc22   2009010968

British Library Cataloguing-in-Publication Data
A catalogue record for this book is available from the British Library.

For information on all Butterworth-Heinemann publications,
visit our Web site at www.elsevierdirect.com

Printed in the United States of America
Transferred to Digital Printing, 2011

To my family

Table of Contents

CHAPTER 8   THE PRODUCT/MARKET MATRIX 181

Preface

Always bear in mind that your own resolution to success is more important than any other one thing.

—Abraham Lincoln

Materials can always be seen as components of engineering systems, forming the basis of new products and leading to innovation. For many companies and particularly those engaged in the aerospace, energy, transportation, construction, "high-tech," or consumer electronic industries, the ability to innovate and produce a stream of new materials-based products is key to the continued growth and profitability of the organization. New materials-based products frequently also serve as a driver to create startup companies. Although, in hindsight, there is always a logical evolution of innovative products, there is no certain method for ensuring innovation. In many cases, recognizing and supporting the "ideas" person and ensuring that the straightjacket of company bureaucracy does not stifle creative instincts is important in establishing an appropriate innovative environment.

Experience has shown that, early on, a focus on a still unclear target is most important and rules and guidelines obtained from business schools may be helpful but are insufficient to ensure success. The role of intuition, pattern thinking, and the ability to make fast decisions are often more important than the application of rational but time-consuming thought processes.

Nevertheless, certain formal processes can be used to encourage the generation of ideas and development of innovative products, and the purpose of this book is to present some of these formalisms in the context of the management of an R&D program in a large diversified company. The content is based on the material presented by me in a course given at the Swiss Federal Institute of Technology in Zurich since 2005. The course was initiated in response to the proposal by Professor Ludwig Gauckler, head of the Laboratory of Nonmetallic Materials, to introduce a focus area on the topic of Materials and Economy and the idea was to take advantage of my experience in managing the corporate R&D program at ABB. It is a frequent criticism of the R&D fraternity that there is insufficient awareness of the costs involved when developing new products, and it was judged that an early introduction to the economics of materials usage would be advantageous.

The book is aimed at graduate and postgraduate students and those early in a research career. In this context, an important feature is the inclusion of case studies from mostly newly established companies in the United States, Canada, Russia, Switzerland, and Finland and my own experience of managing the R&D process to ensure the emergence of new and improved products. Thus, the case studies provide real-world examples of innovation and new product development and bring to life some of the concepts described in the formal chapters. The individual chapters describe the critical phases in product development, including some of the models that can be used to guide the creation and application phases of the dynamics of innovation. Chapters on R&D management, accounting, and market characteristics and requirements provide the reader with the background necessary to understand potential obstacles to be overcome in the introduction of new products.

Most existing management books take a top-down view, based on a statistical evaluation of a large number of case studies, supposedly proving management theories. The approach in this book is bottom up. Although referring to important conclusions from existing management literature, it tries to bring the reader into the real world, where work is done, showing the present and a look into the future. The following points may differentiate this book further.

- The importance of ethical, social, environmental, and sustainability considerations.
- The need to use common sense and intuition.
- The courage to work against all outside resistances and obstacles.
- An analogy between management and mountaineering.

With the emergence of the global economy, the emphasis in the developed countries of Europe and North America in particular is on the production of high value-added products and equipment with a high input of skill and originality in design and production. The information in the book may not ensure an ability to innovate, but it may provide a bridge between academic and professional life and the inspiration to drive the innovative process.

ORIGINS AND ACKNOWLEDGMENTS

The course on Materials and Economy evolved after discussions and input from

- Professor Ludwig Gauckler, who started out his career at the Max Planck Institute, from which he moved to work in an industrial research center of an aluminum company and then became professor of ceramics at the ETHZ, keeping his focus on application-driven basic research.
- Dr. G. Schröder, a leading researcher in metal forming, who held various positions at ABB and later obtained an MBA.
- Dr. W. Hofbauer, an electrical engineer, at one time responsible for ABB's newly created surge arrester business. He taught a somewhat related course to electrical engineers at the ETHZ, providing me with his manuscript, and co-authoring Chapter 7.

From discussions with Gauckler, it quickly became clear that a series of case studies would be of great help to bring real-life experiences to students.

The idea to write this book came during one night in February 2006. I had just completed my first course. I asked the students for their feedback, and they criticized the dominance of too many slides: "Why don't you provide us with some notes as well?"

Many people influenced me during my studies and my subsequent career. Although it is impossible to list them all, I want to give credit to a select few:

- Dr. W. Kaltenegger, who taught me mountain climbing in a constructive way, as if he were my manager.
- The late Prof. F. V. Lenel of Rensselaer Polytechnic Institute, who helped generate my interest in powder metallurgy and its many potential applications.
- Prof. G. Petzow, my first boss at the Max Planck Institute in Stuttgart, for his constructive support.
- Dr. Claus Schüler, my first boss at BBC (later ABB), who led by his interest in my work, including numerous suggestions about my course.
- Dr. Craig Tedmon, CTO at ABB, who taught me the importance of making quick and still good decisions.

Many more people deserve credit. I worked with them, they worked for me, or they were consultants, and I learned a lot from their skills and their patience and enthusiasm: Martin Müller, Erwin Schönfeld, Rolf Lüthi, Dr. C. Corti, Dr. M. Bomford, Dr. C. Buxbaum, Dr. G. Schröder, Dr. P. S. Gilman, Dr. C. R. Boër, Dr. C. Verpoort, Willy Kuhn, Karin Batke, Dr. H. Rydstad, Prof. A. Speiser, Prof. J. Carlsson, Prof. H. Gleiter, Prof. P. Gudmundson, Dr. T. Duerig, Dr. K. Melton, Dr. O. Mercier, Dr. W. Hoffelner, Dr. E. Bakke, Prof. R. Singer, Prof. H. F. Fischmeister, Prof. E. Hornbogen, Denise Riedo, Elisabeth Egli, and Frank Sharp.

As mentioned before, a lot of original information for this book came from case studies. I owe many thanks to Prof. L. Gauckler, Dr. F. Filser, Dr. W. Rieger, S. Jud, Dr. M. Santi, N. Winterberger, P. Keisänen, Prof. J. Pylkkänen, Dr. J. MacDonald, Dr. T. Duerig, Dr. D. Stöckel, B. Zider, Prof. J. C. Palmaz, Dr. H. R. Zeller, Prof. K. Ragaller, Dr. W. Paul, W. Schmidt, M. Hagemeister, and Dr. V. Samarov.

Special thanks go to Dr. T. Gibbons, who spent valuable time to review and improve the manuscript.

Introduction

There are books on materials science, books on processing materials, books on innovation, and books on economics and management. No book combines all these topics from a point of view of the person who starts the innovation process, often a materials engineer but perhaps a physicist, chemist, or anyone with an engineering background.

Innovation is a complex phenomenon. Success requires the combination of many skills, but each situation is so novel that what is needed most often is common sense, the ability to see the bigger picture and quickly decide when to steer into a new direction, how to manage new risks (which appear all the time) and keep looking for newly opened opportunities that may suddenly change the course of the whole process.

This book is an attempt to explain some of the fundamentals, often outside your field of experience but highly important to be successful. You may consider it a personal view of how things are done, which also explains frequent references to the companies Brown Boveri, Asea, and ABB. Much of the information can be found in more detail in the specialized literature, but the purpose of this book is to focus on the essentials, trying to make it interesting by connecting everything to real-world examples. Case studies show what has been learned by actually going through the innovation process or trying to solve a problem. Some of the learning experiences from the case studies are similar, but there is no attempt to develop "theories" around them. With a few exceptions, these case studies do not go into specific technical and financial details, so the focus invariably is at the higher level of decision making and learning experiences. As you will recognize, there is a heavy emphasis on experiences.

ANALYSIS OF AVAILABLE LITERATURE

Numerous textbooks cover the whole range of materials, their properties, methods of processing them, and their applications [1–5]. In the case of mature technologies, the information is compiled in handbooks, often reaching out to cover the interdisciplinary aspects of engineering or economics [6–10]. The value of these books often is based on complete information about the composition, properties, design rules, and applicable economical principles of established materials.

Individual processes, like innovation management, strategic planning, technology roadmapping, financial management, and new ventures, have been covered by many excellent textbooks [11–17].

WHO SHOULD READ THIS BOOK?

To be successful in creating an innovative product, a wide range of skills is necessary. The first glimpse of reality in science comes with the thesis, then working in industry confronts the engineer with the reality of the business world. It is a mistake to believe that an MBA education at such an early stage would help shorten the way to successful innovation. Most likely, the opposite will happen, because the focus on economic and managerial issues is too early. This book is intended to reach both graduates and people in the R&D field, also engineers in small- and medium-sized companies in just about any position.

By definition, any book that offers information about skills that still need to be acquired will be used sooner or later by those who need them. To serve their purpose, such books have to be as interdisciplinary as possible, which often does not happen. This book will not make you a designer of specialized products nor an accountant or a manager. Without realizing it, you would lose interest if too much specialized information were provided, which may just cost you too much time to absorb. The main purpose of this book is to serve as a *bridge from academic, specialized training to professional life*, to widen your horizon and make you aware of what else counts. Experience shows that focused workshops and short-term training courses also do a better job in filling this need than in-depth specialization.

FOLLOWING THE FLOW OF INNOVATION

The book describes all the important aspects that have to be considered, from idea creation to development of an innovative product. These aspects cover the *sources of innovation*, the *people* involved in defining the needs and providing the required knowledge base. To manage innovation, the flow of an *organization* is needed and, as one moves from the initial phase of idea creation into product development, an increasing level of *structure in terms of objectives, strategic planning, and time and cost targets* are required.

The chapters in the book are arranged as much as possible to follow this flow, although due to the multifaceted aspects, a simple time-based approach is not possible.

Chapter 1 describes the first phase, in which ideas evolve from different possible sources of innovation, both outside and within an enterprise. The wide and increasingly interdisciplinary field of materials is described. Twelve case studies, which in the course of this book are discussed in detail, are introduced briefly. The case study CERCON® serves as an example of the convergence of skill sets necessary for a successful innovation.

Chapter 2 introduces the structure and various types of organization of a company. As we will see, there has been a gradual shift away from traditional mechanistic to organic organizations. Five possible ways of departmentalization of companies are described. The process of vision, mission, objectives, and strategic planning is the tool to reach product and business targets.

Chapter 3 focuses on the first part of the creation/application spectrum. The creation phase encompasses research and development. Over time, the well-known linear model of R&D has been replaced by a more-open two-dimensional view, a mix of basic and market-driven applied research. The dynamics of innovation is described by a model exhibiting interdependent rates of product and process innovation over time. Existing innovations invariably experience replacement by a new one within a company or a disruptive process coming in from outside a company. Various models and tools to follow the innovation process are discussed. The case study ZnO varistors addresses several issues: in-house R&D in a technology-oriented but still fairly hierarchically organized company and licensing the same technology by a more market-oriented company.

Chapter 4 deals with application, the second part of the creation/application spectrum. We learn that we have to think backwards from an anticipated product design and define product requirements or specifications for the material to be developed. This includes making the right choice of production engineering and manufacturing. Two case studies address the issue. The first one, on isothermal forging of Ti-alloy impellers, teaches how to compare the manufacturing costs of two existing processes and a new one, still under development. This case also shows the importance of early critical interaction among the various functions involved—R&D, engineering, production, and marketing—which was not done properly here and led to the ultimate failure of this project. The second case study deals with the successful implementation of design-software for hot-isostatic pressing of precision components, overcoming a multitude of initial barriers by relentless interaction among all the potential players.

Chapter 5 describes the challenges involved in managing R&D technology, the first function involved in dealing with new ideas. Many of the comments apply to other functions as well, such as portfolio management, financial commitments, and human aspects needed to successfully support innovation. The case study on managing several corporate research labs at ABB shows you in hindsight, that continuous improvement is needed to keep up to date with worldwide progress in management and be open to new ideas at any time.

Chapter 6 teaches you to understand that any function, R&D included, has to understand the processes and work flows necessary to move forward fast and efficiently. We continue to use ABB as a case study. Initially, PIPE, a new Lotus Notes software program, was developed to continuously manage the work flow of research projects, making it possible to accept new ideas at any time. The Stage-Gate process is an important step forward to integrate all the functions in a large global organization in the decision-making process. Knowledge management, although a bit overemphasized when it became known, shows a new way to distinguish between tacit and explicit knowledge and how it can affect management thinking.

Chapter 7, on the financial management of a company, addresses two major issues. Financial accounting teaches you which tools are available to look at a company's financial status at different points of time: the balance sheet pinpoints financial status at the end of an accounting period—where are we now? The profit and loss account or the income statement shows the results of a company's operations over a period of time in hindsight—where have we been? In budgeting, the same terms are used, but this time to take a look into the future—where are we going? Managerial accounting and investment decisions are the most important tools to be used when developing a product or operating a business.

Chapters 8 and 9 teach you the various ways to develop or grow a business: market penetration for a given product, new product development within an established market, and diversification, where both the product and the market are unknown. Several case studies show that smaller companies, especially start-ups, are more willing to take big risks to succeed with an innovation. What has to be understood, however, is that the likelihood to fail with a new venture is always very large as well.

Chapter 10 shines some light on the human aspects of management. Many tools, also called *psychometric instruments*, measure our capabilities and preferences. Some of them are discussed in more detail. You will learn why you think and act your way and why others may think and act in a totally different way. This allows you to view leadership issues, creativity, and human relations from a new angle. A special emphasis is placed on the importance of intuition as opposed to exclusive rational decision making, as some of the case studies have shown beyond any doubt.

REFERENCES

[1] R.W. Cahn, P. Haasen, E.J. Kramer, Materials Science and Technology: A Comprehensive Treatment, VCH, New York, 1999.

[2] D.W. Richerson, Modern Ceramic Engineering: Properties, Processing, and Use in Design, CRC Press, Boca Raton, FL, 2006.

[3] J.F. Shackelford, Introduction to Materials Science for Engineers, second ed., Macmillan, New York, 1988.

[4] T.A. Osswald, G. Menges, Materials Science of Polymers for Engineers, Hanser-Gardner Publications, Cincinnati, OH, 2003.

[5] W.D. Callister, Materials Science and Engineering—An Introduction, John Wiley & Sons, New York, 1997.

[6] G.S. Brady, H.R. Clauser, J.A. Vaccari, Materials Handbook, McGraw-Hill Professional, New York, 2002.

[7] American Society of Metals, Metals Handbook, ASM International, Materials Park, OH, 1990.

[8] N.P. Cheremisinoff, Handbook of Ceramics and Composites, CRC Press, Boca Raton, FL, 1990.

[9] J. Brandrup, E.H. Immergut, E.A. Grulke, A. Abe, D.R. Bloch, Polymer Handbook, John Wiley & Sons, New York, 2003.

[10] Y. Gogotsi, Nanomaterials Handbook, CRC Press, Boca Raton, FL, 2006.

[11] L.V. Shavinina, The International Handbook of Innovation, Elsevier, Boston, 2003.

[12] R.G. Cooper, S.J. Edgett, E.J. Kleinschmidt, Portfolio Management for New Products, Da Capo Press, Cambridge, MA, 2001.

[13] J.M. Utterback, Mastering the Dynamics of Innovation: How Companies Can Seize Opportunities, Harvard Business School Press, Boston, 1994.

[14] R.C. Dorf, T.H. Byers, Technology Ventures: From Idea to Enterprise, McGraw-Hill Professional, New York, 2004.

[15] P.F. Drucker, The Essential Drucker, Butterworth–Heinemann, Boston, 2007.

[16] P.F. Drucker, Innovation and Entrepreneurship, Butterworth–Heinemann, Boston, 2007.

[17] C. Barrow, Financial Management for the Small Business, Kogan Page, London, 2001.

Reference Notes

From Idea to Product to Market—The Flow

OBJECTIVES

In the section "Materials Development—The Starting Point," the objective is to give a very brief overview of materials science, one of the main sources of innovation. The field has grown rapidly in the past decades, driven by the need for high-performance alloys for demanding applications and, to some extent, by the new field of nanoscience and the increasing interdisciplinary character of the field, which now includes life sciences as an important new element, and new application areas, such as materials for use in information technology.

The flow of innovation is time dependent, although not predictable. The main objective is to use a graphic metaphor of this flow, showing the many factors to be considered in the interaction of human beings with new ideas. A vast number of contributing factors influence this interaction. Some of them—the source of innovation, the role of people and their involvement—are discussed.

Of many sources of innovation, new knowledge is the main driver for innovation in materials science–related topics. To make success more likely, one criterion has to be fulfilled: the convergence of all possible skill levels. The goal of the case study CERCON® is to demonstrate how the convergence of different skills on the level of idea creation and later on the level of idea processing was the key success factor.

Returning to the flow model, several more steps are required to bring a selected idea to a successful target. Scoping, or the screening of many of the original ideas, leads to a focus on one idea, which then must be brought to a certain point, where an informed decision can be made regarding predetermined quality objectives and predefined cost and time targets.

From the two options, careful detailed analysis of the possible variants and fast decision making according to experience levels, the one taking the least time is the one to be favored.

Once the idea has been selected and persons identified to drive the product development process, lots of emphasis must be devoted to understanding all the important processes and ensuring that the level of competence is understood and can be increased in a planned way.

REALITY IN INNOVATION

Innovation typically follows an invention, the creation of a new idea. In the context of materials development, this term is multifaceted. In the most narrow sense it implies a new idea that leads to a new, improved material; but quickly we see that we have to expand the scope to include new ideas for processing the material, new product ideas. We can go further and consider that, for a given set of materials, processes and product innovation can include new marketing methods, better services, or new business models. The term *innovation* is often misused or misunderstood. To many managers, innovation is the exclusive territory of R&D departments and inventors. In fact, innovation should be persistent and systematic, part of a company's culture and processes. Lots of theories have been developed to put innovation into consistent processes. The reality is that there are no hard and reproducible rules to govern innovation, although plenty of available tools help the innovator perform special tasks. To sum it up, tools are important to have, to help to steer you, but they alone are insufficient to steer your course.

Materials are mostly part of very complex systems, which themselves are difficult to optimize, be it functionally or economically. On the one hand, a material property can be life limiting in the product; on the other hand, the material may still be only a tiny gear in a huge gearbox. Take as an example the role of materials used in turbine blades as a part of a bigger system, a turbine, which in itself is part of an even bigger system, a power plant or an airplane. Optimizing performance or cost gives different challenges for the materials engineer. Demands on the system may change, even during the development process, in a sometimes unpredictable way. The more informal and down to earth the process is to find the main influencing parameters, the higher the likelihood of success. Still, statistics about failures in innovation show abundantly that the risks are very high.

So, the problems to solve are these:

How to increase the interdisciplinary skill set?
How to be most flexible in adjusting to unforeseen and unforeseeable changes?

MATERIALS SELECTION IN PRODUCT DESIGN

Mike Ashby [1], over the years, compiled data on most available materials and shows them in a host of property charts (see Figure 1.1, as an example). At least originally, these were really aimed at materials selection, knowing the critical design.

The plot shows the materials attributes for a database of approximately 100 materials. Materials families (polymers, foams, metals, etc.) are identified by

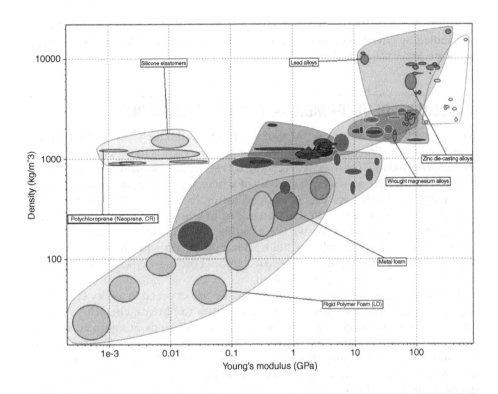

FIGURE 1.1 Example of a property chart using Ashby's CES software.

the larger bubbles. This image was created using Mike Ashby's own CES Selector software and data from Granta Design [2].

The intriguing effect of these plots is that all materials can be grouped into distinct areas, allowing a higher-level comparison among them, like type of atomic forces—again, these are data of well-researched materials. The main driver behind these investigations was the development of software for the designer, who has to select the right material for the right application [3–6]. In more recent studies, the scope has been expanded to the realm of investment methodology for materials, providing tools for materials consultants and investors [7, 8]. These tools are as close to simulating reality as one can expect. The tool for investment methodology for materials [7] is based on the assumption that a material is useful in an application if the balance between its technical, functional, and economic attributes is favorable.

When it comes to starting from a really new material that is still in the research stage (high-temperature superconductors, special ceramics, nanotechnology, or shape memory alloys, to name just a few), we are faced with an abundance of unanswered questions, which never will be answered by any theory or tool.

So, to use Peter Drucker's expression of *convergence* [9], a range of skills has to be handled by either one person or a team of specialists who know how to communicate and exchange views and opinions.

MATERIALS DEVELOPMENT—THE STARTING POINT

We start with a very short overview of a field that is infinitely vast and continues to grow in many directions. The purpose of any overview is to collect all the available information and organize it according to a set of rules equally applied to all of it. The result can be an in-depth analysis of where we stand in our search of new knowledge and which are the areas open for more research. We can also get a comparative insight how different categories of materials were developed and brought to an application. Or we are exposed to completely new areas or new combinations of interdisciplinary fields. What do we do with the result? It is an illusion to think that a top-down approach alone is the most efficient way to solve a problem, leading to a new application.

It is the same with studies at a university. We are exposed in a very systematic way to a large number of fields, from mathematics to physics, chemistry, analytical techniques to materials science. Now, we have all kinds of information gathered, and we are looking for a job. Looking for a job in most cases means entering uncharted territory. The easiest way is to stay at the university or an affiliated research organization and continue to work in the same field in which you started your thesis. Or, you look at companies who gather to meet the graduates because they need persons with a certain combination of skill sets. But most of the time, you have to do the searching yourself, and very often the only questions you ask are: Do they need new materials, and will I be able to add value? When you end up in one of the companies of your choice, you discover that the expectations go far beyond what you learned, and only with an open and fresh mind will you be able to confront the challenges. In my experience, the top-down approach rarely works, but picking up opportunities, as they come along—the bottom-up approach—reflects reality better. In other words, you need to work in an environment where the "bottom" is encouraged to voice its views. The art is to recognize this, be able to analyze quickly which kinds of skills are required to master the job, then enter a learning phase. You also learn that you need not acquire all the skills yourself; you have to find experts and team up with them.

As a researcher in an industrial company, you may be keen to present your results at large international conferences, with two objectives. One is to compare your knowledge in your area of specialty with more knowledge people have available at the time; the second is to expose yourself to completely different knowledge, from fields other than your specialty, at first out of curiosity, but often finding out later that you need that knowledge to enter new territory.

It is in this larger context that we begin with a very gross overview of the various fields in materials science. One good source of information, if one can read beyond the political goals, is the *European White Book on Fundamental Research in Materials Science* [10]. Many of the following comments are based on information from this book, and they can be seen in a global context as well.

Materials science, by definition, is an interdisciplinary research field, which covers the areas of physics and chemistry of the various classes of materials but also engineering aspects, when it comes to the design, manufacturing, and application areas. Life science has become an area of new interest in materials science, mostly due to the evolution of nanotechnology. A simple, but somewhat outdated way of classifying materials is to put them into three categories:

- Metallic materials.
- Nonmetallic and ceramic materials.
- Polymeric materials.

The invention of the atomic force microscope, in 1986, opened another dimension on the atomic scale, nanoscience, which will have a long-term impact on industrial development comparable only to information technology.

Both the excursion into nanoscale dimensions and the widening of the interdisciplinary aspects of material science, as well as a wide range of applications, ask for a more detailed way of classifying materials, often focusing on the functional aspects of materials.

Materials Systems (and Related Case Studies)

Although the reader should be encouraged to think of materials in terms of applications, a few remarks on the wide range of materials should be made here.

Materials can be divided into two broad groups: inorganic and organic materials. Inorganic materials are composed of non-carbon-based materials, such as metals, ceramics, and metalloids (e.g., silicon and germanium). Organic materials are based on carbon compounds and include both living tissue and synthetic polymers. Another way to classify materials is according to new technological concepts (nanomaterials) or their main application areas: materials for information technology, catalytic materials, and many more (Figure 1.2).

To connect different materials to product and business development, a number of *case studies* have been worked out and are listed along with the various material classes. Later in the book, each of the case studies is described in greater detail. Each case study is different, and each has its special learning points. You are invited to think through each case study yourself: How would I have acted? What would I have done differently—for better or worse? What do I learn from one case study that I can apply to later ones?

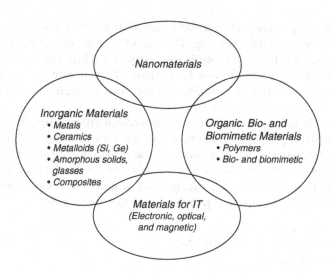

FIGURE 1.2 Overview of material systems according to various classification criteria.

Inorganic Materials

Metals and Metallic Alloys Including Composites

These materials still constitute the backbone of mechanical engineering industries: automotive industry, aerospace, power generation, and electrical engineering, to name just a few of these industries. The drivers for R&D can be summarized as follows:

- Higher strength but lower cost (Fe-based materials).
- High strength at lower specific weight (Al and Mg alloys).
- Resistance against oxidation, corrosion, creep, thermomechanical fatigue, and the like, at increased temperatures (superalloys).
- Improved physical properties in materials for special applications (hard and soft magnetic materials, electrical resistance and electrical contact materials, shape memory alloys, hard metals).
- New and advanced processing techniques (forging, casting, powder metallurgy, coatings, bonding).

Usually specialized handbooks (e.g., [11]) give an almost complete overview of the state of the art. Modeling and computer simulation of many materials have reached a level where the performance of novel and advanced conventional structural and functional materials can be understood and predicted, thus bridging the gap between materials and design. Modeling and computer simulation also play an increasing role in understanding, modifying, and predicting the synthesis of materials, the processing, the resulting microstructures and the properties, which are a function of these. Two case studies address this issue: the case study on iso-thermal forging of Ti impeller wheels and the case study on hot isostatic pressing.

Case Study on Isothermal Forging of Ti Impeller Wheels.   Here, a team of material scientists, manufacturing engineers, and specialists in process modeling worked together developing successfully process models predicting every stage of the forging process as well as mechanical properties of the forging [12]. The term process model refers to a mathematical model which has to be developed to a level at which it can quantitatively describe the essential characteristics of a pro-cess. When implemented as a computer program, it permits the stepwise simula-tion of the process. The practical aim of a computer process model is always the generation of quantitative statements concerning the process—faster, cheaper and more extensive than can be obtained by laboratory experiments or tests during actual fabrication; it replaces the trial and error approach with "right first time" and "net shape."

Process models have to be seen as the technology modules within a CAE (computer-aided engineering)/CAD (computer-aided design) system, which permit an engineering treatment of the process-related problems.

Case Study on Hot Isostatic Pressing.   This is an example of the convergence of the knowledge of how to do complex mathematical calculations and the newly avail-able high-speed computing to predict the right shape of a preform, which leads to a high-precision final shape in hot isostatic pressing. The case also shows how, following consistent guidelines, a business was developed that started with process modeling and went on to one of the leading ventures making extremely complex components.

Two more case studies deal with metallic materials: case study on nitinol devices and the case study on lanthanum oxide–molybdenum cathodes.

Case Study on Nitinol Devices.   Shape memory alloys, when discovered, immedi-ately offered a wide range of novel properties, thus generating a lot of ideas in many places of the world. This case study, however, shows that the path to a suc-cessful product and business based on stents can be very lengthy, and a lot of obstacles, the least of which are technical, have to be overcome.

Case Study Lanthanum-oxide Molybdenum Cathodes.   Here, starting from a vision that cathodes for radio transmitter valves containing La would have a higher elec-tron emission than thoriated tungsten cathodes, challenges had to be met concerning the lifetime and the fabrication of molybdenum wires containing a fine dispersion of lanthanum-oxide. Skills at different locations and in different fields had to be combined to make this development possible, although the proj-ect failed for other technical and commercial reasons.

Ceramic Materials
Ceramic materials are considered to have the largest range of physical properties of all known materials. Technology areas that benefit from advanced ceramics are

energy technology, transportation, medical technology, information technology, and manufacturing technology.

G. Fantozzi et al. [13] list some of the new ceramic materials, which deserve attention in future research:

- *Inert or active bioceramics* for application in the medical and health fields.
- *Hard and superhard ceramics*, such as abrasives in grinding wheels and cutting tools for metalworking.
- High-temperature corrosion of particle filters.
- Ceramic membranes for liquid or gas hyperfiltration.
- *Electroceramics*, one of the widest application areas, with activities ongoing in the fields of ion-conducting ceramics for batteries and sensors, electrical insulators in integrated circuits, semiconductors (sensors), and superconductors.
- *Ferroelectric ceramics* for applications such as capacitors, sensors, piezo-electric transducers, electrooptical devices, and thermistors (resistors to measure temperature changes).
- *Ferromagnetic ceramics* for use in the electronics industry.
- Miniaturization for electroceramic components and devices, which is an ongoing challenge.
- Particle filters for diesel engines with better corrosion resistance.
- Ceramic brakes in transportation with improved performance and reduced lifetime cost.
- Components for gas turbines.
- Solid-oxide fuel cells.
- Improvement in the thermomechanical and chemical behavior of refractories used in power generation.
- More efficient thermal barrier coatings in gas turbines.
- Fiber-reinforced ceramic composites for high-performance thermomechanical applications.
- Improved understanding and prediction of density variations during pressing and sintering to obtain near-net shape components.

Six case studies deal with a large variety of ceramic materials and application areas.

Case Study on ZnO Varistors. Two approaches to develop a product, in-house R&D and licensing, are described and compared. The case study also gives a short historical view of the different cultures at two companies that would later merge.

Case Study on Liquid Crystal Displays at Brown Boveri. Although Brown Boveri was a partner in pioneering the LCD (liquid crystal display) effect and successful in setting up a manufacturing facility, followed by a joint venture with Philips, the lack of market knowledge, combined with the problem of high-cost manufacturing in Switzerland, meant that the business went to Japan. Still, Brown Boveri/ABB succeeded in profiting from licensing the technology worldwide.

Case Study on Rapid Prototyping of Zirconia (CERCON®). A new machine concept to produce high-precision dental implants is described. To make it successful, several skills had to converge, those from a visionary dentist and a leading researcher in high-strength ceramics. One engineer, combining most of the diverse skill sets (convergence) required to develop the product, led the project, bringing the idea to a successful conclusion within a very short time.

Case Study on Metoxit. Starting out in a large aluminum company, many new ideas in ceramics were evaluated, gradually shifting the emphasis to highest-strength ceramics and applications in the biomedical field, which further led to the creation of a new venture. One highlight is the use of intuition and common sense as opposed to applying only rational thinking.

Case Study on High-Temperature Superconductors. Intensive research worldwide followed the discovery of a new high-temperature superconductor (HTSC). Applying technology push, impressive results were shown by demonstrator units of power devices by large electrical engineering companies. A startup, American Superconductors, was the only one to have the foresight and focus to develop a successful business by teaming up with companies that were potential users of the technology.

Case Study on Day4Energy. Photovoltaic cells have been around for a long time, and many solutions exist. By the congruence of a novel idea, created by a Russian physicist, reducing the series resistance between silicon and the current conductor, and a Canadian entrepreneur, a new business was started, originally with the idea of selling it off to the highest bidder, but then leading to the rapid buildup of a large, successful manufacturing company.

Amorphous Solids—Glasses
There is still plenty of room for discovering new vitreous materials and a need for a better understanding of the glass formation mechanisms and the modeling of glass structures.

Structural Composite Materials
Composites are created by bringing together two or more disparate materials, resulting in superior properties or improved functionality. At present the main systems under investigation are metal matrix and ceramic matrix composites. It is expected that biological materials will be used as well in composites. The main areas of application are in transportation, electronic packaging, power transmission, aerospace, and sports technology.

Organic, Biomaterials, and Biomimetic Materials

Polymeric Materials
The field of polymeric materials is broad and covers the whole range from basic synthetic chemistry to process engineering. Drivers of innovation are

- New synthesis methods, applying
  □ New mechanisms of polymerization and combination of polymerization methods.
  □ Nanopatterning via block copolymer self-assembly.
  □ Selective solubilization of low-molecular-weight compounds.
  □ High purity synthesis.
- Novel architectures and materials combinations:
  □ Composite materials for greater strength and functionality.
  □ Nanostructured materials based on polymers.
- Functional polymers:
  □ For optics and electronics, such as light emitting diodes, displays, sensors, batteries.
  □ Membranes for fuel cells and batteries.
  □ High-performance polymers (polyimides, fluoropolymers, etc.).
  □ Organic-inorganic hybrid structures.

Biomaterials and Biomimetic Materials

Biomaterials and biomimetic materials also are known as *biocompatible materials*. The most important issues deal with interaction between the biomaterial and cells. They are used as sensors in tissue responses, as bone cement in bone-bonding systems, in joint replacements, in skin repair devices, in systems facilitating localized drug delivery, in dental implants for tooth fixation, and in contact lenses, to list just a few.

Biomimetic materials are materials not made by living organisms but having compositions and properties similar to those made by living organisms. An example is the calcium hydroxylapatite coating found on many artificial hips used as a bone replacement, which allows for easier attachment of the implant to the living bone. Biomimetic materials are used not only for medical purposes but as structural, smart, and functional materials as well.

Nanomaterials

The first exposure to the idea of materials properties in the nano dimension came in 1973, when Professor H. Gleiter was visiting the Brown Boveri Corporate Research Center to discuss with the scientists some of their research projects. At that time, he was well known through his work on modeling the structure of grain boundaries. Once he said, "one day there will be totally new materials with the structure of grain boundaries, which are at the order of nanometers in dimension." Nordmann [14] independently addressed Gleiter's pioneering role in shaping the first visions of this class of materials.

A huge amount of government-funded research has dealt with the field of nanotechnology, and it is rapidly evolving as a source of an increasing number of innovations. Taking the "bottom-up" approach in generating materials, as compared with the traditional "top-down" approach, opened a completely new viewpoint in creating atomic arrangements with completely new properties. Nanomaterials are designed and used today in a variety of forms, such as nano-powders, colloids, thin films and coatings, multilayers, and laterally structured systems. Colloid science has given rise to many materials that may be useful in nanotechnology, such as carbon nanotubes and other fullerenes and various nanoparticles and nanorods. Nanoscale materials can also be used for bulk applications and medical applications.

Key areas of applications are seen in IT and communication, biomimetics, health care, quantum computing, chemistry, energy and sustainability, and nano-machines, but we can take it for granted that a large number of unplanned innovations will surface in the next few years.

Two case studies describe the successful application of these new classes of materials, demonstrating finally that a large number of products are beginning to emerge from previous basic research efforts.

Case Study on NanoSphere®

A mid-size Swiss textile company, in the need to survive by new high-tech innovations but without substantial internal R&D facilities, succeeded developing dirt-repelling clothing by using the so-called Lotus effect, but delegating the research to larger, well-qualified outside R&D facilities. The required skills were the competence of the person in the company supervising the external research and the flexibility of the company to develop a new product in a short time.

Case Study on Amroy

Amroy evolved as a fast startup from the Finnish Science Centre in Jyväskylä, where a group of scientists and engineers were considering several possible applications of nanoscience. Almost by coincidence, they discovered that ultrasonic cleaning of CNTs (carbon nanotubes) improved their atomic bonds to a matrix of epoxy resin. Although coming up with CNT-reinforced polymer was an obvious idea, it took this special process discovery to make the idea work and move it into the product development stage. Partnering with established companies having experience in polymer processing and access to various markets helped grow this new business rapidly.

Electronic, Optical, and Magnetic Materials

Electronic and photonic materials encompass a broad range of substances from semiconductors, such as doped silicon, germanium, and GaAs, to electronic and ionic conductors made of metals, polymers, and ceramics. Here is a list of various application fields:

- Materials for Si-MOS (nano) electronics.
- Materials for high-technology electronics, solar cells, sensors, and the like.
- Materials for spin electronics.
- Organic materials for LEDs, displays, and molecular electronics.

- Materials for lighting.
- Materials for optical communication systems.

In magnetic materials, new discoveries in colossal magnetoresistance, magnetic multilayers, magnetic properties of thin films, magnetic nanowires, and magnetooptic layers will lead to many new applications.

INNOVATION FLOW

Before going into the specific processes of product or business creation, we deal with the concept of innovation. Many definitions and books go into great detail about what innovation is. In reality, the term is often used interchangeably with other ones like *invention* and *improvement*. Considering that all these terms are used interchangeably in most companies, let us use the definition that is easiest to understand: *The successful introduction and exploitation of new ideas.* This definition implies an underlying process of making improvements by introducing something new, a creative idea that has been realized.

In the context of this book we should consider two aspects:

1. New materials and manufacturing processes may lead to both incremental and radical changes to products, processes, or services. The goal of innovation is to solve a problem. From an economical point of view, innovations must increase the value for either the producer or the customer.

2. In the organizational context, innovations can be related to increases in performance and improvements through higher productivity, better efficiencies and quality, but also to an increase in market share.

Since innovations can and often do fail, it is important to understand that they carry a certain risk. To minimize this risk, a careful balance between product and process innovations should be sought.

With this still quite wide definition of *innovation*, we can see it taking place in all parts of an organization. Innovations can occur in

- Products
- Processes
- Marketing
- Supply chain
- Services
- Organization
- Business model.

We can distinguish between two major types of innovation: incremental and disruptive innovations.

Incremental innovation is a small step forward in technology from the known to the unknown. Generally, due to the small degree of uncertainty, the risks involved are quite low and can be easily managed. It is therefore not surprising that most established companies pursue incremental innovation.

Disruptive, breakthrough, or radical innovation may involve an entirely new product or service. Because of the higher risks that come along with it, failures are quite common, but the rewards in case of success can be overwhelming. It is understandable and normal that proposals for disruptive innovation meet with skepticism, opposition, and many questions relating to the cost. There is often considerable uncertainty about the future: Will the product meet all specifications? Will the customer want it?

Innovation by companies can be achieved in many ways. Breakthrough and more radical innovations often originate from a formal R&D department, whereas incremental innovations evolve most likely from the operational business units. One has to be aware, though, that there are always exceptions to these trends.

Important Aspects in the Flow of Innovation

Innovation—now defined as the generation, completion, and transformation of ideas into products leading to a business—is a process that evolves over time. It can also be shown graphically as a narrowing down of the alternatives, finding the right focus and aiming successfully at a target, which often may be different from the one originally envisaged. But, many aspects have to be considered and have to be seen in a holistic way:

- What are the sources of innovation? Where do ideas come from?
- Who can be involved in the generation of ideas?
- Which is the right idea from an abundance of those created?
- What is the relative importance of a rule-based rational approach compared with one based on intuition and common sense?
- How can one find the right focus for the quality objective, cost, and time?

One way to visualize the complexity is shown in Figure 1.3. Ideas are generated from different sources of innovation, by different types of people at different locations. They then undergo a screening and evaluation process, which narrows them down to a few with a higher likelihood of success. Still, the course is never clearly predictable, and decisions have to be made all the time to change or readjust the course.

Where Do Ideas Come From? Sources of Innovation

Innovation generally stems from the goal-oriented search for opportunities. Peter Drucker [15] identified that opportunities for innovation exist both within and outside a company or industry. Opportunities internal to a company include unexpected events, incongruities in processes or between expectations and results, process needs, and changes in the marketplace or industry structure. Opportunities external to a company include demographic changes, changes in perception, and new knowledge.

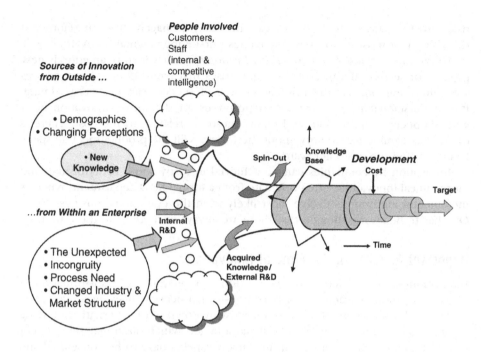

FIGURE 1.3 From source to launch, the flow of ideas in innovation.

Sources Within the Enterprise

1. **Unexpected events**. Unexpected events can be failures as well as successes. The unexpected failure of isothermal forging of turbocharger titanium impellers was because a better design of the impeller came up and the market required more and different types of turbochargers but in lesser quantities. However, this unexpected failure opened the new opportunity of flexible manufacturing of more complex-shaped components, which no longer were needed in the quantities needed to economically operate a forging plant.

2. **Incongruities**. Incongruities result from a difference between perception and reality. Toyota was able to capitalize on a growing number of consumers wanting to lower the carbon-dioxide emissions of their cars and willing to pay a premium for a hybrid car. The opposite has happened so far with fuel cells, where companies such as Daimler, and many others too, invested hundreds of millions of dollars to develop fuel-cell-powered cars, which would require not just a fairly low-cost product but also an entirely new infrastructure for providing hydrogen or related fuels to the customer.

3. **Process needs**. Process need innovations are those created to support some other process or product. Switching from glass plates and films to digital processing in photography totally changed the way cameras are used now. Digital processed photographs can easily be sent to practically anyone

in the world via the Internet. Likewise, the development of the ATM (automatic teller machine) freed bank tellers from performing many routine functions, such as cashing checks, and improved both efficiency and profit margins for banks.

4. **Changes in industry structure or market structure**. Such changes catch everyone unaware. Industry structures change in response to growth and changes in the marketplace. Before the advent of the personal computer, manufacturers of large mainframe computers dominated the computer industry. With the adoption of the personal computer and the advent of the laptop computer, the composition of computer sales and marketing changed dramatically. The long-expected change in the automotive market toward more fuel-efficient models had to await the sudden rise of oil prices, at least in the United States.

Changes Outside the Enterprise

5. **Demographics**. An increasing source of innovation is coming from countries with rapidly rising populations, such as India and China, and from changes in the makeup of the population. China has become a testing ground for some of the most advanced new technologies in transportation and power transmission because it still greatly lacks infrastructure, making it cheaper to install new technologies, which would be too expensive in more developed countries with an existing old-technology infrastructure.

6. **Changes in perception**. Although effective initiations are still too low on a worldwide scale, the threat of global warming has started many initiative moves, leading to lower emissions by using new products: a boom in photovoltaics, the introduction of low-power incandescent lamps, wind power, and many more.

7. **New knowledge**. New knowledge revolutionized the field of materials science in the past 20 years, starting with nanotechnology, which helped enforce the integration of several disciplines: physics, chemistry, materials science, and life sciences. Most case studies in this book center on this source.

Where to Get Information?

Ideas are generated by people, and these can be at very different places.

Direct customer contact is often the best source for innovation. This contact helps to understand their needs better than information obtained through in-house information gathering and helps to quickly come up with clear specifications of a new product.

Partners can be anyone with special knowledge outside the company's knowledge base.

Markets provide often specialized information on a higher level, making it easier to focus on the right course.

Internal staff with complementary skills is a must, especially when going from materials to materials systems, continuously widening the scope of the task.

The *Internet* has become a knowledge base of continuously increasing importance. It helps to access all knowledge bases and look for connections among them. *Extranet* can be understood as a private intranet mapped onto the Internet or some other transmission system not accessible to the general public but managed by one company's administrators.

Internal databases are very important sources of information, especially if they reach back a few years.

A *patent search* is one of the best ways to obtain more complete, detailed information about new technologies, information that could not be obtained by visiting conferences or reading publications.

Competitive intelligence helps us to understand what different thought processes led to a new product idea.

Academia presents its information at conferences or in published papers.

In the *discovery phase*, decisions are made on the information that leads to various types of ideas: product ideas, platform ideas or technology ideas.

New Knowledge as a Source of Innovation

New knowledge is the predominant source used in entrepreneurship. While it gets most of the publicity, most of public money, and attracts the interest of scientists worldwide, it also contains both very high risks and very high rewards in case of success. Although it seems to be the obvious path to take for scientists, from creating a new idea to a successful business, history shows that the path can often be convoluted and very long, with many barriers to overcome. Time spans of 20 to 40 years for materials-based businesses are not uncommon. Creating NDC (Nitinol Devices Company) and Metoxit (producing high-strength zirconia biomedical applications) each took 20 years. Fuel cells were discovered in 1837, and although billions have been spent on R&D worldwide, an undefined number of years still is needed until the commercial success of one of the many product ideas is reached.

Knowledge-based innovation requires a careful analysis of all the necessary factors. Beyond the ones that deal with the knowledge, social and, importantly, economic factors have to be put into the equation. It is important to know exactly where one wants to go: The whole system around a material-based idea has to be fully understood, the right market focus has to be found. Both parameters need careful consideration to come up with a clear strategic position. Too often, a new business is established by a person knowledgeable in a particular field. To be successful, the knowledge-based innovator needs to learn and practice entrepreneurial management—or find someone to do it. Many new ventures were started by a technically oriented person, who after several years had to make room for a better qualified general manager. Time is against the innovator in science- and technology-based innovations, and the window of opportunity is often quite small. As soon as a new idea emerges, others are motivated to follow in the same direction, hoping they will come up with an improved idea and bring it to market faster.

Also, the time has to be right. Very often an innovation fails, because it was developed too soon. Further, control of intellectual property rights (IPR) is of key importance. A good idea is not worth much unless it is protected by IPR. In numerous cases, someone came up with a new idea, had it protected by a patent, and later had to win against large companies that commercialized the invention.

Principles of Successful Innovation

One has to be realistic. Many innovations are not developed by following a clear, well-organized path in a systematic manner. Many of the best case studies show the importance of knowing intuitively how to maintain focus on the goal and be ready to make corrections in the course as progress is made and unexpected obstacles show up. MBA studies give you a set of tools, which you may need from time to time, but you will not escape the responsibility to make decisions on your own. This is why each innovation has its own history. It cannot be replicated, it cannot be taught, and therefore it cannot be learned. Successful innovators have to be conservative; they must not be constantly shying away from risk, although they should have good knowledge about them. They must seize opportunities as they come and move fast.

An important requirement for success is what Drucker calls *convergence*. Knowledge-based innovations are multidisciplinary in nature. Therefore, different skill sets are required, which have to come together (converge). Most of the time this happens when a group of people with different skills get together in a team. But, in a few cases, convergence is possible in one person, as the following case study shows.

Case Study. CERCON®: Convergence of Skill Sets (Personal Interview with F. Filser)

"One other key characteristic of knowledge-based innovations—and a truly unique one—is that they are almost never based on one factor but on the convergence of different kinds of knowledge, not all of them scientific and technological."

—Peter Drucker,
Innovation and Entrepreneurship (1993) [15]

Objective of the Project

The objective of the project was to introduce high-strength zirconia into dental applications and develop as a product a rapid prototyping process [16,17]. Several persons with very different skill sets came together to make the development of a new product from the materials laboratory and its introduction into the market a success:

- A visionary dentist/professor (P. Schärer at University of Zurich), who could define the specifications and requirements for the new product and critically supervise and accompany the project.
- An applications-driven professor (L. J. Gauckler), who had led in establishing a know-how base of special ceramics, in particular zirconia.

- A champion (F. Filser), who combined and developed all the skills needed as the project evolved, in the fields of
  - ☐ Mechanical engineering.
  - ☐ Information technology.
  - ☐ Materials and production technology.
  - ☐ Marketing research.
  - ☐ Intellectual property rights and licensing contracting.
- A motivated partner company (DeguDent), providing up to 50 staff specialists to speed up the demonstration and technology transfer phase.

Making it all work required, in addition, the skill to communicate among different knowledge bases and, if necessary, go a step further and take over some of the work usually required from the partner.

Original Idea and History of R&D Leading to CERCON®

The first idea came in 1990, when cooperation with a German dental company was initiated, applying hot-isostatic pressing to zirconia. A new idea—machining presintered ceramics then sintering it to full density—was pursued from 1995 to 2000 with another company, which, however, was developing a competing product (filled polymer). Frank Filser and Peter Kocher came up with a draft machine concept in 1997 (Figure 1.4) but needed to work with a company that had the skills to develop a professional CAM (computer-aided manufacturing) machine.

Mechanical Engineering

No market data were provided to Filser, neither could he find a competent partner to design the machine. Finally, in 1999, a contract was signed with DeguDent, and within six weeks a first machine prototype was produced (Figure 1.5), to a great extent due to his knowledge in mechanical engineering.

Information Technology

An important part of the manufacturing system is the gathering and storage of data about the shape of the tooth to be manufactured. These data have to be stored on a microchip that is inserted into the machine, which then executes the direct ceramic machining (DCM) process. Understanding the basics of IT and integrating this knowledge into the design of the machine were important parts of the overall knowledge required.

Market Research

Since no valuable market information had been provided, Filser took it upon himself to personally conduct the market research, making thousands of phone calls: "for some time I had the largest telephone bill at the university." The market investigation led to the following conclusions: Dental labs are small enterprises with a restricted budget per year and limited resources. Direct contact with the decision maker, usually the boss of the lab, is required to get useful comments regarding an investment.

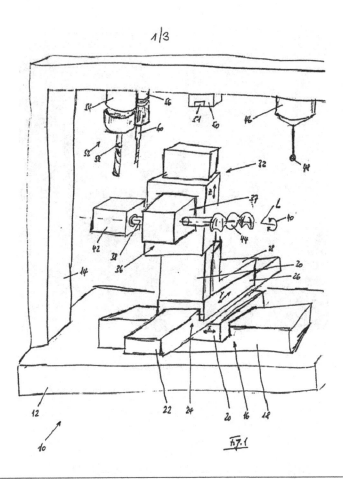

FIGURE 1.4 Draft machine concept for machining presintered zirconia [18].

The *main learning experiences* from this market research were the following:

- Get business intelligence.
- Benchmark the system against the competition.
- Establish a unique selling position—the product has to be both better and less expensive.

Materials Research

A wealth of experience about high-strength ceramics existed at the materials lab of the ETHZ (Swiss Federal Institute of Technology, Zurich). Combining this knowledge with the market experience of P. Schärer, an overview of all existing alternative materials was provided and zirconia was established as the highest-strength material for this application (Figure 1.6).

FIGURE 1.5 Single-purpose prototype machine with optical sensor for dental technicians.

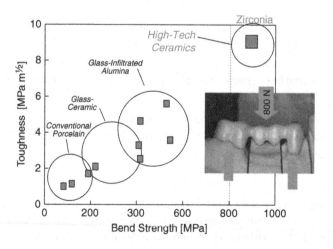

FIGURE 1.6 Properties of ceramics in dentistry.

Patent Strategy

One can learn a lot by continuously analyzing existing patents. Patents, by definition, contain all the relevant details, which may not be found in a scientific or technical publication. Nevertheless, one also has to build up his own patent family. To keep costs down, focus on the few countries most representative of the company's market. It is important to file an intellectual property claim as early as possible, through a patent attorney with good international experience, who can put your claim in the proper order.

Working with Suppliers

The designed machine consists of many parts and components that will be made by suppliers. It is therefore important to make good, open contacts with potential suppliers. The best way to get the right information is to treat them as partners, give them the right information about the expected number of parts you want to purchase, and ask for quotations in the required numbers, delivery times, conditions, quality control, and so forth. As we never know the future exactly, it is of vital importance to develop various scenarios and discuss them with the suppliers.

Important Steps

1. **Mind mapping and SWOT.** To put the whole process, which combined many overlapping disciplines, into one coherent picture, the tool of mind mapping [19] was used to overcome one of the main human problems. Many people sit in a "silo" that they find difficult to leave. A scientist is familiar with addressing the question of unknown science, which can run the danger of putting too much emphasis on finding new knowledge rather than pursuing the main goal of developing a marketable product. Figure 1.7 is a mind map addressing the topics of science, production, application, market, and time management. It can be seen as a checklist of all the things to be considered. Correctly defining the specifications and target costs requires learning about the competition, applying SWOT analysis (dealing with strengths, weaknesses, opportunities, and threats), and knowing their prices.

2. **Finding the right cost of the production unit.** When comparing offers from different suppliers, each one came up with a different number. Most likely, each supplier company is structured differently; each one thus has a different cost accounting system. Only by comparing different offers can we, by iteration, approach the optimum cost for new production equipment.

3. **Finding all processing steps.** The actual number of processing steps is much larger than that perceived by a materials engineer, because it includes "trivial" steps such as packaging (see Table 1.1). To know them all is important, because they all require time and cost money.

Transfer to Market

The project had evolved from the university, although it involved cooperation with potential suppliers. The next step was to find the right business model to transfer

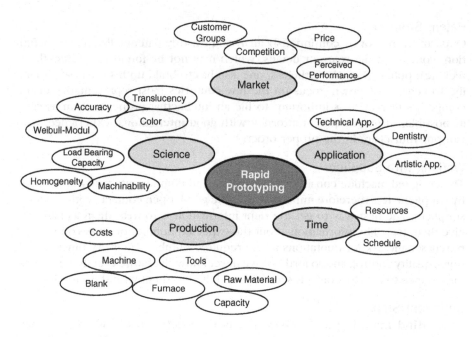

FIGURE 1.7 Mind map for rapid prototyping of CERCON®.

Table 1.1 Example of the Complexity of Listing All Processing Steps

Processing Steps

1. Purchase and store powder
2. Fill in powder
3. Cold press
4. Machine green compact
    a. Machine along axis
    b. Trim
5. Presinter
6. Fill container with parts to be sintered
7. Turn on furnace
8. Remove sintered parts from container
9. Cut to right lengths
10. Perform quality control
    a. Check weight, length, diameter
    b. Report data
    c. Evaluate statistical data
11. Attach bar code
12. Attach adaptor
13. Pack
14. Deliver the product

the product to the market. To find the right one, all potential alternatives had to be known and assessed. Let us assume the name CERAG for the company that would market the product. We can then distinguish among these four scenarios.

1. CERAG produces and sells the system to the end user.
2. CERAG sells the system but buys all systems components from third parties.
3. CERAG produces the system and sells only via a few distributors.
4. CERAG sells licenses to companies (an existing or new player on the market) to produce and sell the system.

We then weigh the arguments for and against the two main options: establishing a startup company or licensing to an established player (Table 1.2).

The learning from this is that the best solution is to first create your own spin-off then look for an existing player as a licensee.

The next step is to come up with a *contract* with the licensee, consisting of two phases:

- Evaluation, which can take 1.5 years; it also contains an up-front payment for the know-how.
- Licensing after completion of the first phase and containing a second payment plus royalties.

The goal of licensing should be commercialization, not the "shelving" of the technology. Both parties should have the right to exit the contract. The contract has to contain

- A list of the patents and a description of the know-how.
- The company's business plan and the project plan.

Table 1.2 The Two Main Options

	Pro	Con
Startup	Be in control in establishing and running the business Get the entire profit	Access and service costs of many small dental technical labs Conservative clients trust only well-established brands and technologies Danger to be bought by a big player
Licensing	Less risk Established market players have brand name, access to clients, distribution channels	Only small part of profit for inventors Willingness to make invention work

To prevent misunderstandings, paragraphs have to be written in clear wording, or otherwise be rejected. Upcoming problems can best be resolved in meetings and in a climate of partnership with the contractor. Let the contractor become the owner of the project, accepting that it may go off course a few times.

Conclusions and Learning Points

By the end of 2002, the first year of market entry of CERCON® systems (Germany, Austria, Switzerland, United States), 400 systems and more than 200,000 units (blanks) were sold; the total revenues were €30 million.

By the end of 2005, more than 1500 systems were sold worldwide; more than 20 companies were using the presintered zirconia technology, building up the zirconia market.

To Frank Filser, the whole project caused lots of extra work, but it also gave him satisfaction to see his ideas on the market. Lots of contacts to industry and lots of additional knowledge could be generated. It helps to think about commercialization in an early stage, which helps to anticipate the steps required to make it happen.

Scoping—Selecting the Few Winning Ideas

Scoping is a quick, preliminary investigation of each project. Without defining the scope of the project we cannot estimate either cost or time. Sometimes lack of proper communication can lead to frequent revisions of scopes. This directly spirals costs and disturbs the schedule of the project leading to losses.

Using a few selection parameters or a simple screening tool, most of the work can be achieved quickly by desk research. A simple series of questions can be used to address the following:

Customer benefit: Does the idea offer a clear benefit?
Business benefit: Can the product based on this idea be sold profitably?
Technical feasibility: Can this idea be realized given the current technology?
Time feasibility: Can this idea be implemented on schedule?

Importance of Speed in Decision Making

Speed in decision making and determination of who makes the decision are most important. Thus, from the two possible approaches—rational decision making vs. the use of intuition—the second one is the one to choose, but it requires experience and certain skills. Eisenhardt [20] summarized some of the important criteria to be met to be successful in making fast strategic decisions in a fast-paced environment:

- Availability of real-time information.
- Number of simultaneous alternatives considered.
- Presence and participation of experienced people.

- Skill in conflict resolution.
- Integration among strategic decisions and tactical plans.

Finding the Right Focus: Time, Cost, Objectives

Once the right idea has been selected, a process has to be initiated to define clear goals addressing these three questions:

How long will it take (time)?
How much will it cost (cost)?
What is the quality objective of the project (objective)?

All these issues are addressed later in the book, which itself is structured to resemble the flow of ideas.

Right Processes and Right Competencies—Important Pieces of a Puzzle

Learning about sources of innovation, the network of people involved, and setting up a strategy is not enough. We have to make sure that all important competencies for the key processes are highly developed and everything fits together well, as in a puzzle (Figure 1.8). These are

- Knowledge of competence development and management.
- Knowledge of R&D.
- Knowledge of product development.
- Knowledge of manufacturing.
- Knowledge of marketing.

All processes are important, and they can be considered pieces of a puzzle put together, with an emphasis on the competencies continuously growing to assure success of developing a product and a business.

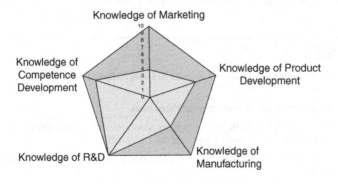

FIGURE 1.8 Right processes and right competencies.

SUMMARY

The introduction differentiates this book from more specialized ones. It combines topics of materials science, of materials processing, of innovation, economics and management, thereby acting as a bridge for the academically trained specialist to the industrial environment.

The section "Materials Development—The Starting Point" covers the vast and still growing field of materials science. Areas such as nanotechnology and the addition of life sciences are fairly recent developments, which, however, have a huge impact on the content of innovation. The case studies are listed here briefly, showing which area of materials science each addresses. The purpose of the case studies, which is discussed in greater detail in the next chapters, is to explain, by a bottom-up approach, that the rules to be followed are hardly firm, but that there is a constant need to reconsider the original targets and be ready to readjust the course quickly.

A graphical metaphor showing the flow of innovation from its creation to a finished product serves as a guide through the various chapters of this book, as the various phases of innovation are discussed. The flow starts with a discussion of Peter Drucker's seven sources of innovation. Looking at the case studies covered in this book, new knowledge is singled out as the most important source for innovation in the materials field. Another important point to make innovation a success—the convergence of skill sets—is discussed in the case study CERCON®. The case study shows that excellence in materials science and dentistry is not enough to make the development of this new product successful. The product is a device permitting precision machining of presintered ceramic preforms, which are subsequently sintered to full density to dental implant components. Skills in materials engineering, mechanical engineering, process optimization, information technology, marketing, and communication with both suppliers and end customers had to converge—in this case study, mainly in a single person—to bring the project to a quick success.

REFERENCES

[1] M.F. Ashby, Materials Selection in Mechanical Design, Butterworth–Heinemann, Boston, 2005.

[2] Granta Design, available at www.grantadesign.com.

[3] National Materials Advisory Board, Computer-Aided Materials Selection during Structural Design, National Academy Press, Washington, DC, 1995.

[4] C. Yann, P. Siarry, Multiobjective Optimization: Principles and Case Studies, Springer-Verlag, New York, 2004.

[5] R. Bernard, Multicriteria Methodology for Decision Aiding, Springer-Verlag, New York, 1996.

[6] K. Paul Yoon, Ching-Lai Hwang, Multiple Attribute Decision Making: An Introduction, SAGE Publications, Thousand Oaks, CA, 1995.

[7] E.M.A. Maine, M.F. Ashby, An investment methodology for materials, Mater. Des. 23 (2002) 297–306.

[8] E. Maine, D. Probert, M. Ashby, Investing in new materials: A tool for technology managers, Technovation 25 (2005) 15–23.

[9] P.F. Drucker, The Essential Drucker, Butterworth–Heinemann, Boston, 2007.

[10] European White Book on Fundamental Research in Materials Science, Max-Planck-Institut für Metallforschung, Stuttgart, 2001.

[11] American Society of Metals, Metals Handbook, ASM International, Materials Park, OH, 1990.

[12] C.R. Boër, N. Rebelo, H. Rydstad, G. Schröder, Process Modelling of Metal Forming and Thermomechanical Treatment, Springer Verlag, Berlin, 1986.

[13] G. Fantozzi, D. Rouby, J. Chevalier, R. Reynaud, Advanced Ceramic Materials: Summary of Possible Applications, in: European White Book on Fundamental Research in Materials Science, Max-Planck-Institut für Metallforschung, Stuttgart, 2001, p. 32.

[14] A. Nordmann, J. Schummer, A. Schwarz, Nanotechnologien im Kontext, Akademische Verlagsgesellschaft, Berlin, 2006.

[15] P.F. Drucker, Innovation and Entrepreneurship, Butterworth–Heinemann, Boston, 2007.

[16] F. Filser, P. Kocher, F. Weibel, H. Luethy, P. Schaerer, L.J. Gauckler, Zuverlässigkeit und Festigkeit vollkeramischen Zahnersatzes hergestellt im DCM-Verfahren, Int. J. Comput. Dent. 4 (2001) 89.

[17] L.J. Gauckler, P. Kocher, F. Filser, Rapid Manufacturing of High-Tech Ceramics. A Case Study for Dental Application, in: L. Jan, E. Jean-Pierre (Ed.), Proceedings of the Second International Conference on Shaping of Ceramics, Mol, Belgium, 2002, pp. 259–264.

[18] F. Filser, L. Gauckler, P. Kocher, H. Luethy, P. Schaerer, Machine Tool for the Production of Base Structures for False Teeth, European Patent EP1246581.

[19] J. Nast, Idea Mapping: How to Access Your Hidden Brain Power, Learn Faster, Remember More, and Achieve Success In Business, John Wiley & Sons, New York, 2006.

[20] K.M. Eisenhardt, Making Fast Strategic Decisions in High-Velocity Environments, Acad. Manage. J. 32 (1989) 543–576.

An Introduction to Company Structure and Organization

OBJECTIVES

After reading this chapter you will be able to understand

- The changing economic environment in which we live, and the changing demands on the qualification of engineers.
- The various forms of organizing a company, the advantages and disadvantages.
- The purpose of mission, vision, strategy, and objectives in finding the direction in which to develop a business.

ECONOMIC ENVIRONMENT

We live in a time when understanding technology is becoming increasingly important. Research and development, although sometimes not acknowledged, is often the most challenging function in an enterprise and critical to its survival. Thus, in high-tech companies, engineers play a significant role, and they carry a substantial responsibility for securing its future. This also increases the risk of making mistakes, if the engineer lacks economic knowledge about the enterprise and its relation to world economics.

To be successful as an engineer, it is therefore increasingly important to be capable of dealing with economic issues, be they in development or management. One of the objectives of this book, using the example of the surge arrester (ZnO varistors), is to show the impact of technical decisions on the economic success of a company. The arguments used can be generalized, though, and applied to other products and business segments.

Worldwide, many needs are still unsatisfied. If we define *demand* as a need for which satisfaction can be bought, then we quickly see that only a fraction of the world's population has the purchasing power to satisfy the demand. The question thus arises: How can we generate purchasing power in those markets with a still

29

existing but not yet a satisfied demand? If no answer can be found, then worldwide overcapacities in manufacturing will evolve. The consequence of this will be the continuous need to reduce manufacturing capacities or transfer them to lower-cost countries, and only those companies will survive that have the lowest marginal costs.

Innovation will be of increasing importance in saturated markets. Innovation, in this context, means not only product and process innovations but includes marketing and organizational innovations. For the engineer, this requires widening his or her horizon beyond the field of science and technology.

Major changes are caused by the drive to automate and use electronics wherever this is possible and economically viable. The energy content of products also continuously will be decreased, while the knowledge content increases. New technologies increasingly will affect more than one industrial segment.

The consequence of increased productivity has been and will be a continuous reduction in the number of classical industrial workers, who more and more will be replaced by machines. Examples of such developments are numerous (steel industry, aluminum industry, automotive industries, and electrical engineering industries, to name just a few). The cost of labor as a percentage of the total cost will continue to decrease, and low-cost labor will be replaced increasingly by fewer, higher-skilled experts.

Contrary to predictions from the 1950s through the mid-1980s, persistent shortages of nonfuel minerals have not occurred, despite increasing consumption, and world reserves have increased. New methods have been developed to recycle raw materials. A typical product of the 20th century is the car. Here, the percentage of recyclable raw materials is larger than 40%. The issue of recycling is gaining in importance and is increasingly considered a purchasing argument.

Major products of the 21st century will be the semiconductor chip and, increasingly, the use of silicon in photovoltaic cells. Silicon is processed from sand, which exists in abundance. Still, there is an increasing shortage of processed silicon.

Indicators show that the prices of material resources will increase, due to the rapid economic growth in countries such as China and India.

There is, however, a shortage of the resource *knowledge*, which becomes of value only when being used by knowledge management, a discipline that has received increasing attention in the past few years.

The demographic situation has many implications. Not only will the world's population increase to a level of 8.5 billion people within the next 15 to 20 years, the average age of the population will also increase, generating problems to finance pensions. Careers will change rapidly in the future. There will be fewer hierarchical promotions, because many challenging positions are already occupied by highly qualified young experts, and there is a general trend away from traditional structures of organization to more organic ones. Today's career is defined as assuming responsibility for a wider range of tasks, so an ever-increasing knowledge base is a requirement for the careers of the next generation of engineers.

OBJECTIVES AND ORGANIZATION OF COMPANIES— ESTABLISHED ONES AND THOSE TO BE ESTABLISHED

When asked the question, What is a company's main responsibility? the most frequent answer today would be that it has to return money to the shareholders and maximize its profit. In such a business model, the human being is only a replaceable part in the manufacturing process. In reality, the company has many purposes, depending on the interested group. Foremost the company has an obligation toward

- Its *customers*, to satisfy a demand.
- Its *employees*, to offer them satisfactory work and a secure income.
- The *investors*, to produce a return on their invested capital.
- *Suppliers*, to whom the company offers a market.
- *Society* for creating jobs and satisfying demands.
- The *state*, to pay taxes.

In the words of David Packard (1948), cofounder of the Hewlett Packard Corporation, the "purpose of a company is not to make money; it makes money in order to be able to do what it's really about it—to make a contribution" [1]. Peter F. Drucker (interview, September 1999) phrased it similarly [2] "Believe me—no financial man has ever understood a business. That is, because financial people believe that companies make money—but companies do not make money, they make shoes," and then, "When I was a financial analyst myself, it took me a very long time to learn that companies do not make money and that one does not start by analyzing financial statements but analyzes the business. The one is at the end and not at the beginning."

In a good company, this can be translated into the positive pressure employees feel to make a contribution and produce something that is worth more to the customers than it costs to make. Profit can be seen as a measure of how well the corporation can satisfy the needs of the customer:

$$\text{Profit} = \text{Income from sales} - \text{Sum of all costs}$$

Structure of a Company

Most of the business literature evolves around the legal aspects, pertaining both to ownership and the question, Who runs the company? On more practical terms, when one sees a company as an entity that wants to organize itself to make a contribution and do its job well, it requires a clear set of rules of responsibilities, work processes, and so forth.

The first step is to define all the tasks (functions) to reach the business objectives. Typically these functions include market analysis, research and development, marketing, purchasing, production, sales, and financial management.

Then, the tasks are delegated to functional areas. Each function has a clearly defined task and has to see itself as part of a bigger picture. Interfunctional communication is the key to the success in reaching the business objectives.

The formal way to have a system with a coordinated division of labor is to set up an organization. In small- and medium-size enterprises the number of recurring tasks is often so small that standardization makes little sense. Such companies often have no structured organization. Just think of the one-person company. Here, one person alone has to do all the tasks of all the functions. When an entrepreneurial business becomes bigger, one person cannot do all the work without assistance so the entrepreneur needs more and more people to share the work. Because of the small size, interpersonal contact is possible among different functions and hierarchies. Most communication is done by one-to-one conversations. Larger companies require more bureaucratic forms of organization. Here, we can either speak of the *structural organization* (or organizational structure) or the *organic organization*.

In a second step, various tasks are delegated to functional units, called (bottom-up) *departments*, *main departments*, *business units*, *business areas*, *divisions*, *segments*, and the like. The terms can vary from company to company, but the idea is the same: bringing together people and groups in the most efficient way to secure communication and execution of the work.

The structuring of tasks, competencies, and responsibilities requires organizing the company into units. The following have to be taken care of:

- Distribution of tasks to units.
- Hierarchical ordering into units and subunits that lead and execute the tasks.
- Leadership issues, where competencies and responsibilities are sorted out.

In any type of organization, employees' responsibilities typically are defined by what they do and to whom they report; for managers, the responsibility is determined by who reports to them.

Since people frequently change their jobs and positions, a time-independent way of describing these positions and their relationships, an *organizational chart*, is created. The best organizational structure for any organization depends on many factors, including

- The work it does.
- Its size in terms of employees and revenue.
- The geographic dispersion of its facilities.
- The degree to which it is diversified across markets.

From a Traditional (Mechanistic) to an Organic Organization

At the beginning of the 20th century, as industry was shifting from job-shop manufacturing to mass production, lots of thinking was applied to these new systems. Ideas evolved that the best way to structure organizations for the greatest efficiency and productivity was very much like a machine [3,4]. Even before this, Weber [4] had concluded that, when societies embrace capitalism, bureaucracy is the inevitable result. Management's thought was originally influenced by Weber's ideas of bureaucracy, where power is ascribed to positions rather than to the individuals holding them. Taylor [3] took the scientific approach, trying to find the "one best way" to accomplish a task using scientifically determined studies of time and motion. Fayol [5] focused on the concept of unity within the chain of command, authority, discipline, task specialization, and other aspects of organizational power and job separation. This created the context for vertically structured organizations characterized by distinct job classifications and top-down authority structures, or what became known as the traditional organizational structure, also called *mechanical organization.*

Job specialization, hierarchical reporting, and the subordination of individual interests to the higher goals of the organization together led to organizations arranged by functional departments, with order and discipline maintained by rules, regulations, and standard operating procedures. Even though the differences among organizations are enormous, they share many similarities that enable them to be classified. One widely used classification is the twofold system (mechanistic versus organic forms of organizational structure) developed by Tom Burns and G. M. Stalker [6]. Mechanistic systems are best suited to stable conditions, whereas organic systems are appropriate to changing conditions. Intermediate stages may occur between the extremes, and firms may operate within both systems at the same time.

Traditional Organizational Structure

The structure of every organization is unique in some respects, but all organizational structures develop or are consciously designed to enable the organization to accomplish its work. Typically, the structure of an organization evolves as the organization grows and changes over time.

Generally, managers have to make four basic decisions as they design an organizational structure, although they may not be explicitly aware of these decisions.

1. The organization's work must be divided into specific jobs. This is referred to as *division of labor.* Work process requirements and employee skill levels determine the degree of specialization. To maximize productivity, supervisors match employee skill level with task requirements. Supervisors perform *work flow analysis* to examine how work creates or adds value to the ongoing processes in an organization. Work flow analysis looks at how work moves from the customer or demand source through the organization to the point at which the work leaves the organization as a product or service to meet customer demand. Major performance breakthroughs can be achieved through *business process reengineering,* a fundamental rethinking and radical redesign of business processes to achieve dramatic

improvements in costs, quality, service, and speed. Business process reengineering uses work flow analysis to identify jobs that can be eliminated or recombined to improve company performance.

2. The jobs must be grouped in some way, which is called *departmentalization*.

3. The number of people and jobs to be grouped together must be decided. This is related to the number of people to be managed by one person, or the *span of control*, the number of employees reporting to a single manager.

4. The way *decision-making authority* is to be *distributed* must be determined.

In making each of these design decisions, a range of choices is possible. At one end of the spectrum, jobs are highly specialized, with employees performing a narrow range of activities; while at the other end of the spectrum, employees perform a variety of tasks. In traditional mechanistic structures, there is a tendency to increase task specialization as the organization grows larger. In grouping jobs into departments, the manager must decide the basis on which to group them. The most common basis is by function. For example, all scientists and engineers can be grouped into an R&D or engineering department, all accounting jobs in the organization can be grouped into a financial department, and so on. The degree to which authority is distributed throughout the organization can vary as well, but traditionally structured organizations typically place final decision-making authority to those highest in the vertically structured hierarchy. Final decisions usually are made by top management. The traditional model of organizational structure is thus characterized by high job specialization, functional departments, narrow spans of control, and centralized authority. Such a structure has been referred to as *traditional*, *classical*, *formal*, *mechanistic*, or *command and control*. Not surprisingly, to take the example of large Swiss companies around 1970, prior military experience was assumed to be the best qualification for top management positions.

A structure formed by choices at the opposite end of the spectrum for each design decision is called *unstructured*, *informal*, or *organic*.

The traditional model of organizational structure is easily represented in graphical form by an organizational chart. This is a hierarchical or pyramidal structure with an executive at the top, a small number of senior managers under the president, and several layers of management below this, with the majority of employees at the bottom of the pyramid. The number of management layers depends largely on the size of the organization. The jobs in the traditional organizational structure usually are grouped by function into departments, such as engineering, production, accounting, sales, human resources, and the like. There are traditionally three types of organizational structures: functional organization, divisional organization, and matrix organization.

Functional Organization

The functional organization is a structure in which authority rests with the functional heads. Each function can be further divided into departments (Figure 2.1).

Depending on the level of authority connected with the organization we distinguish between line organization and staff-line organization.

Line Organization The line structure is defined by its clear chain of command, with final approval on decisions affecting the company still coming down from the top. Line structures are most often used in small companies and are, by nature, fairly informal and involve few departments.

Staff-line Organization This organization is similar to the line organization, but contains additional staff functions, with the purpose of decreasing the workload of the functional leaders and making them independent of their subordinates. Examples would be human resource management and financial management.

The advantages are increased quality in the decision-making process through the contribution of specialists and reduction of workload for both leaders and subordinates. The disadvantages are the possible misuse of power through the staff specialists, an increase in overhead, accompanied by higher costs and slower decision processes.

Summing up, what are the advantages of a functional structure?

- It has a clear division of tasks and competences.
- In a clear hierarchy, coordination is the task of top management.
- Fast decision making is always possible.

What are the disadvantages?

- Too many decisions may be made by the top management. Especially in large organizations, this means that the gap between the knowledge base at the working level and at the management level becomes too big.
- Horizontal coordination of functional tasks is difficult.
- Little rewards for cooperation among various groups make this coordination even less attractive.
- This all leads to a loss of information.

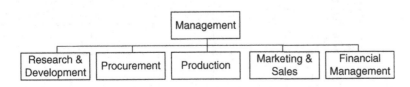

FIGURE 2.1 Functional organization.

Divisional Organization

In a divisional organization, corporate divisions operate as relatively autonomous businesses under the larger corporate umbrella. Divisional structures are made up of self-contained strategic business units that each produce a single product. For example, ABB's divisions or segments now include Power Technologies and Automation Technologies. The central headquarters, focusing on results, coordinates and controls the activities and provides support services between divisions. Functional departments accomplish division goals. A weakness, however, could be a tendency to sometimes duplicate activities among divisions.

Matrix Organization

Some organizations find that none of the aforementioned structures meet their needs. One approach that attempts to overcome the inadequacies is the matrix structure, which is the combination of two or more different structures. This type of organization is used by companies that have to respond quickly and flexibly to more complex, dynamic situations. Each employee is assigned two bosses in two hierarchies.

Functional organization commonly is combined with product groups on a project basis. For example, a product group wants to develop an addition to its line; for this project, it obtains personnel from functional departments, such as research, engineering, production, and marketing. The personnel then work under the manager of the product group for the duration of the project, which can vary greatly. These personnel are responsible to two managers, as shown in Figure 2.2.

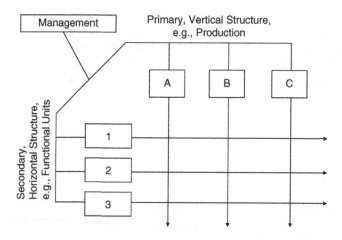

FIGURE 2.2 Matrix organization.

One advantage of a matrix structure is that it facilitates the use of highly specialized staff and equipment. Rather than duplicating functions, as would be done in a simple product department structure, resources are shared as needed. In some cases, highly specialized staff members may divide their time among more than one project. In addition, maintaining functional departments promotes functional expertise, while at the same time working in project groups with experts from other functions encourages cross-fertilization of ideas. There is less stress on management because of delegation of some of the tasks to other functions. Finally, working together in effective teams helps increase the personal employee satisfaction.

The disadvantages of a matrix organization arise from the dual reporting structure. This may cause problems about assignment of priorities within the matrix, power struggles between managers, and confusion among the employees about who is actually the boss. In the end, the person with control over the money might be the key person.

The organization's top management must take particular care to establish proper procedures for the development of projects and keep communication channels clear so that potential conflicts do not arise and obstruct organizational functioning. In addition to the product/function matrix, other bases can be related in a matrix. Large multinational corporations that use a matrix structure most commonly combine product groups with geographic units. Product managers have global responsibility for the development, manufacturing, and distribution of their own product or service line, while managers of geographic regions have responsibility for the success of the business in their regions.

A good example of such a matrix organization was ABB after its creation in 1988. ABB promoted as one of its most famous slogans, Think globally, act locally. Very quickly almost independent businesses were set up in various European countries and in all major countries in the Americas, Africa, and Asia (China, India). A matrix-type of structure helped eliminate the competition between diverse locations, which had been known to exist in some of the companies of Brown Boveri, one of the two founding companies of ABB. Due to the very large number of legal business units, though, communication was not always easy and was addressed by partial centralization. Figure 2.3 shows a high-level organization chart of ABB in 1998.

Organic Organization

Except for the matrix organization, all the structures just described focus on the vertical organization, that is, who reports to whom, who has responsibility and authority for what parts of the organization, and so on. Such vertical integration is sometimes necessary but may be a hindrance in rapidly changing environments. A detailed organizational chart of a large corporation structured on the traditional model would show many layers of managers; decision making flows vertically up and down the layers.

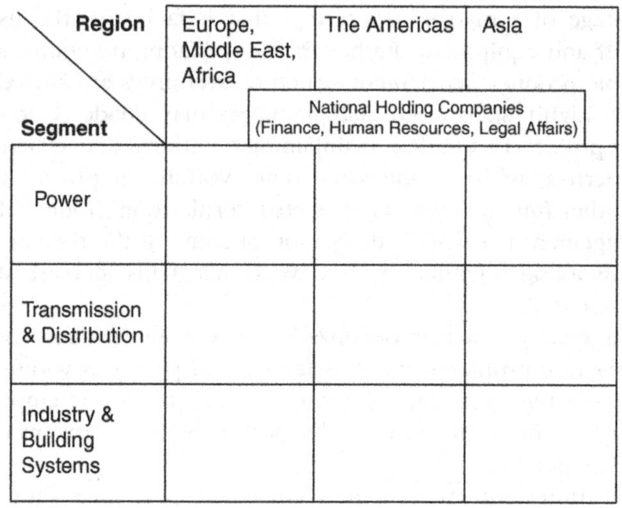

Region ⟍ Segment	Europe, Middle East, Africa	The Americas	Asia
	National Holding Companies (Finance, Human Resources, Legal Affairs)		
Power			
Transmission & Distribution			
Industry & Building Systems			

FIGURE 2.3 ABB matrix organization, 1998.

In any organization, the people and functions do not operate completely independently. To a greater or lesser degree, all parts of the organization need each other. One approach is to flatten the organization, to develop the horizontal connections and deemphasize vertical reporting relationships. At times, this involves simply eliminating layers of middle management. In a virtual sense, technology is another means of flattening the organization. The use of computer networks and software designed to facilitate group work within an organization can speed up communication and decision making. Even more effective is the use of intranets to make company information readily accessible throughout the organization. The rapid rise of such technology has made virtual organizations and boundaryless organizations possible, where managers, technicians, suppliers, distributors, and customers connect digitally rather than physically.

A different perspective on the issue of interdependence can be seen by comparing the organic model of organization with the mechanistic model. The traditional, mechanistic structure is characterized as highly complex because of its emphasis on job specialization, highly formalized emphasis on definite procedures and protocols, and centralized authority and accountability. Yet, despite the advantages of coordination that these structures present, they may hinder tasks that are interdependent. In contrast, the organic model of organization is relatively simple [6]. Organic organizations are characterized by

- Decentralization of authority.
- Flexible, broadly defined jobs instead of job specialization.
- Interdependence among employees and units.

- Free flow of communication throughout the organization.
- Employee initiative.
- Relatively few and broadly defined rules, procedures, and processes.
- Shared decision-making and goal-setting processes at all levels.

A key issue in organic organizations is determining who has the knowledge, expertise, or skills required to identify opportunities or find solution to problems. Rather than assuming that top management is the source of all knowledge and wisdom, organic organizations assume that various people in the organization may have the right insights or capabilities.

Diversity of information and experience is often the key to creative responses to ill-defined, complex problems and opportunities.

A common way that modern business organizations move toward the organic model is by the implementation of various kinds of teams. Some organizations establish self-directed work teams as the basic production group. Examples would be production cells in a manufacturing firm. At other organizational levels, cross-functional teams may be established, either on an ad hoc basis (e.g., for problem solving) or on a permanent basis as the regular means of conducting the organization's work. Part of the impetus toward the organic model is the belief that this kind of structure is more effective for employee motivation.

The advantage of more organic forms of organization is that they are appropriate in turbulent, unpredictable environments and for nonroutine tasks and technologies [7]. Organizations coping with uncertainty need to find the right, effective, and timely responses to outside challenges.

Mixing Styles

The organic organization is not entirely without hierarchy or formalized rules, procedures, and processes. Structural parameters, even if loosely or broadly defined, are necessary to prevent the chaos that would result from absolute decentralization (i.e., where everyone in the organization is completely free to decide what he or she wants to do or not do). As an example of such structural parameters, while employees of Minnesota Mining and Manufacturing (3M) are encouraged to take the initiative in suggesting new products and seeking support from others in the organization, new product teams must still meet specific financial measures at each stage of product development. Nonetheless, the real control is found through constant interaction among peers and the normative rules that develop informally among them.

It is not always necessary or possible for an entire organization to be organic. Some units, such as research and development departments, may benefit from an organic structure because they face an unstable environment. Units that have a more stable environment, such as administrative departments, may favor a mechanistic structure.

The structures of organic organizations are informal, fluid, and constantly changing to identify and develop responses to new problems and opportunities.

Authority and responsibility shifts from one situation to another. Groups are established, complete their work, and disband. In organic organizations, diminished emphasis is placed on superior/subordinate roles in favor of dispersed initiative. Roles, tasks, and responsibilities are not limited by rigid, vertical boundaries of hierarchy for decision making, communication, coordination, and control.

Basis for Departmentalization

Departmentalization is a method of subdividing work and workers into separate organizational units that take responsibility for completing particular tasks. Many organizations group jobs in various ways in different parts of the organization, but the basis that is used at the highest level plays a fundamental role in shaping the organization. There are five traditional ways of how to group activities (Figure 2.4): functional departmentalization, geographic departmentalization,

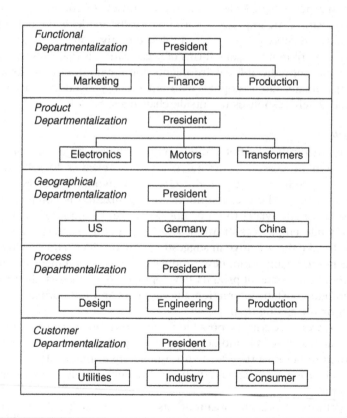

FIGURE 2.4 Five ways of grouping activities by departmentalization.

product departmentalization, customer or market departmentalization, and process departmentalization.

Functional Departmentalization

Every organization of a given type must perform certain jobs to do its work. For example, key functions of a manufacturing company include production, purchasing, marketing, accounting, and human resources. Using such functions as the basis for structuring the organization may, in some instances, have the advantage of efficiency. Grouping jobs that require the same knowledge, skills, and resources allows them to be done efficiently and promotes the development of greater expertise. A disadvantage of functional groupings may be that people with the same skills and knowledge will develop a narrow departmental focus and have difficulty appreciating any other view of what is important to the organization; in this case, organizational goals may be sacrificed in favor of departmental goals. In addition, coordination of work across functional boundaries can become a difficult management challenge, especially as the organization grows in size and spreads to multiple geographical locations. The art is then to find the middle course between centralizing activities between functions and widening the scope of each function.

Geographic Departmentalization

Organizations that are spread over a wide area may find advantages in organizing along geographic lines, so that all the activities performed in a region are managed together. In a large organization, simple physical separation makes centralized coordination more difficult. Also, important characteristics of a region may make it advantageous to promote a local focus. Brown Boveri had a clear geographical departmentalization, but there was often clear competition between some of the divisions in different geographical locations.

Companies that market products globally sometimes adopt a geographical structure. In addition, experience gained in a regional division can be an excellent training for management at higher levels.

Product Departmentalization

Large, diversified companies are often organized according to product. All the activities necessary to produce and market a product or group of similar products are grouped together. For example, ABB used to have business segments in power generation, power transmission and distribution, automation systems, transportation, and oil and gas, to name just a few. In such an arrangement, the top manager of a product group or a segment has considerable autonomy over the operation. The advantage of this type of structure is that the personnel in the group can focus on the particular needs of their product line and become experts in its development, production, and distribution. A disadvantage, at least in terms of larger organizations, can be the duplication of resources. Each product group requires personnel in most of the functional areas, such as finance, marketing, production, and other functions. The top leadership of the organization must decide how much redundancy it can afford. Compromises are possible, where some of the functional areas are partially or completely centralized, as it often happens with R&D laboratories, where costly testing equipment and specialized skill sets can be shared across a range of product departments.

Customer or Market Departmentalization

An organization may find it advantageous to organize according to the types of customers it serves. For example, a distribution company that sells to consumers, government clients, large businesses, and small businesses may decide to base its primary divisions on these different markets. Its personnel can then become proficient in meeting the needs of the different customers. In the same way, an organization that provides services, such as accounting or consulting, may group its personnel according to these types of customers.

Process Departmentalization

Departmentalization by process groups jobs on the basis of product or customer flow. Each process requires particular skills and offers a basis for homogeneous categorizing of work activities. These services are each administered by different departments. This can be defined as a structure in which the work processes are ordered. Let us imagine a task which has to be solved, such as getting a newly developed product to the market for the first time. The R&D department lacks the skill to do this job alone. Cooperation with other functions, such as production, marketing, and sales is needed to succeed. Once the tasks have been decided, the work processes need to be determined to reach the target customer at a predefined time. To make this happen, persons and resources have to be allocated to do the job. An important way to allocate value to managing processes is the value chain concept developed by M. Porter in 1985 [8].

Porter's Value Chain

A value chain is a chain of activities. Products pass through all activities of the chain in order, and at each activity, the product gains some value. A firm can be modeled as a chain of value-creating activities. Porter identified a set of interrelated generic activities common to most companies. The resulting model, known as the *value chain*, is shown in Figure 2.5.

The *primary activities* include inbound logistics, operations (production), outbound logistics, marketing and sales (demand), and services (maintenance). The goal of these activities is to create value that exceeds the cost of providing the product or service.

- **Inbound logistics** includes the receiving, storage, and inventory control of input materials.

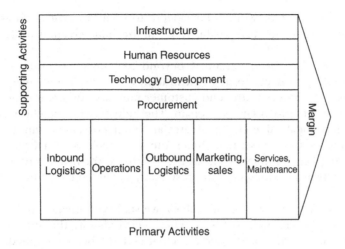

FIGURE 2.5 Porter's value chain.

- **Operations** are the value-creating activities that transform the inputs into the final product.
- **Outbound logistics** are the activities required to get the finished product to the customer, including storage and order fulfillment.
- **Marketing and sales** are those activities associated with the customer to purchase the product, including pricing and advertising.
- **Services and maintenance activities** are those that maintain and enhance the product's value, including customer support and repair services.

Any or all of these activities may be vital in developing a competitive advantage.

The *supporting activities* include administrative infrastructure management, human resource management, information technology, and procurement.

- **Infrastructure management** includes financial operations, strategic planning, quality management, public relations, and the like.
- **Human resource management** encompasses the activities associated with recruiting, development, and compensation of employees.
- **Technology development** includes research and development, process automation, information technology, marketing research, and other technology development required to support the value-chain activities.
- **Procurement** is the function of purchasing materials, machines, advertisement, and services.

Supporting activities are often viewed as overhead, but companies are trying to get competitive advantages there as well, for example by building networks and outsourcing major R&D activities.

The value chain framework is considered to be a powerful analysis tool for strategic planning. Its ultimate goal is to maximize value creation while minimizing costs.

The concept has been extended beyond individual organizations. It can apply to whole supply chains and distribution networks. The delivery of a mix of products and services to the end customer mobilizes different economic factors, each managing its own value chain. The industrywide synchronized interaction of those local value chains creates an extended value chain, termed the *value system*. A value system includes the value chains of a firm's supplier (and their suppliers all the way back), the firm itself, the firm distribution channels, and the firm's customers (and presumably extended to the buyers of their products).

To better correlate activities and values, we start with the generic value chain then map the specific process flows, which helps identify the individual value-creating activities. Once the discrete activities have been identified, linkages between activities where performance or cost affect each other have to be found. Competitive advantages may be obtained by optimizing and coordinating linked activities. One big advantage of a value-chain analysis is that it provides data to make outsourcing (make or buy) decisions.

SETTING UP VISION, MISSION, OBJECTIVES, STRATEGIES

As discussed in the previous section, organizational structures are created in the best possible way, so that a company's work can be performed and its goals can be met. A purpose lies behind the existence of any company, that is why the company was created. Also, companies have to set themselves targets and use tools to establish the right targets and be successful in reaching them. It is the purpose of a set of business processes to link together clear visions of the future with well-defined, step-by-step operations to make it possible to reach goals.

Strategy is a process through which a company makes sense of the world around it, how it is now, and how it may evolve. Because strategy can capture a company's best thinking only at a given point in time, it needs to be revised and refined as people gain new experience and knowledge. Figure 2.6 shows the process flow for *strategic planning*.

This process is most useful in strategic management at the level of individual business units. For large corporations consisting of many business units, a higher-level strategy process, concerned with managing a portfolio of businesses, is more appropriate. Typical corporate-level decisions deal with the questions, which business units should grow, how to allocate resources to various business units, how to find synergies among various business units, and with the issue of selling a business unit or mergers and acquisitions.

Vision

⇓

Mission

⇓

Objectives

⇓

Situation Analysis

⇓

Strategy Formulation

⇓

Implementation

⇓

Control

FIGURE 2.6 Flowchart of the strategic planning process.

Vision

For a company to survive and grow in the long run, there must be an often ambitious wish or dream where it wants to be in the future—a vision. Having a vision helps the company come through setbacks, remembering what the company stands for in the first place. A vision, though, does not contain a path how to get there. This is described by the mission.

Mission

A company's mission is its reason for being. It is often expressed in the form of a mission statement. A mission should be measurable. It projects a company image to customers and describes what the organization as a whole will achieve and the overall timescale for achieving it, but it does not deal with how the end is to be achieved.

While this kind of approach is endemic in most modern organizations, it remains to be seen whether it has any significant impact on the innovation process.

Objectives

Objectives are specific goals the company wants to reach. Objectives should be very ambitious but achievable. They should also be defined in such a way that they can be measured, to allow monitoring its progress and making corrections if necessary.

Situation Analysis

The next step is the collection and compilation of both the internal and external data that can help to delineate the current situation. The external data includes market, economic and competitive intelligence, customer and supplier information, and benchmark data from both inside and outside the directly competitive industry. The results of any survey work and other relevant data are reviewed and discussed to identify the group's present internal *strengths* and *weaknesses* and the external *opportunities* and *threats* (SWOT) that affect the organization's vision and strategies for the future. Figure 2.7 shows how a SWOT analysis is related to the situation analysis.

In the course of the strategic process, starting with a situation analysis, a lot of information is generated. SWOT acts as a filter to single out the most relevant information. *Strength* is an indication where a company can build a competitive advantage, whereas *weaknesses* provide a chance for improvement to become more competitive. *Opportunities* from the external analysis can be translated into new profitable businesses, whereas *threats* are a warning signal for corrective measures.

Strategy Formulation

Now that we know why a company exists and what the mission is, we want to decide how to achieve it. Accomplishing this is the role of a strategy or the strategic plan. A strategy is a fairly high-level look at goals or key milestones that need to be achieved if the mission is to be accomplished. It also explains how the organization as a whole will do it.

The strategy is a set of major steps that are planned. It describes how each step will be reached, who is responsible for it, how long it will take, how much it should cost, and which are the agreed-on measures of success. It is also a method that purposefully focuses a company's resources on specific targets, targets that offer the best prospects for long-term growth, stable profitability, and a competitive edge.

FIGURE 2.7 Deducing SWOT from situation analysis data.

We can simplify it with one sentence: *A strategy is a means of reaching goals*. To reach these goals, we have to answer these five questions:

Where are we going?
When will we get there?
How do we measure progress?
How much can it cost?
Who will do it?

Companies, almost without exception, comprise more than one business unit. A company can then be said to constitute a portfolio (or division, segment, or something else).

We therefore have to be aware of the need to follow a certain sequence in the process of strategy analysis:

Portfolio
Business unit
Functional unit

Developing a strategy can also be seen as a way to come nearer to the goals, which themselves are driven by the vision of the company (Figure 2.8).

This leads to strategy development, covering

Vision
Mission
Objectives or Goals
Strategies
Programs and Projects

FIGURE 2.8 Segmentation of business strategies into functional strategies.

Strategic Planning

The whole purpose of strategy development is to secure a strategic advantage (or competitive edge). To do this, we first have to position the firm with respect to others and the whole industry. Michael Porter [8] argued that a firm's strength falls into one of two categories: cost advantage and differentiation. By further applying these strengths in either a narrow or broad scope, we can introduce focus as a third generic strategy. The term *generic* is chosen because these strategies are independent of firms and industry. Figure 2.9 illustrates Porter's generic strategies.

Cost Leadership Strategy

This strategy calls for becoming the low-cost producer in an industry for a given quality. Cost advantages can be achieved by

- Improving process efficiencies.
- Finding unique access to lower-cost materials or processes.
- Outsourcing or vertical integration

To successfully implement such a strategy a firm needs several internal strengths:

- Access to capital for investments in new production equipment.
- Capability to design products for efficient manufacturing.
- Expertise in manufacturing process engineering.

At the same time a company must be aware of the risk that other firms may follow a similar strategy or even leapfrog by developing a new, more cost-efficient product design or manufacturing process.

Target Scope	Advantage	
	Low Cost	Product Uniqueness
Broad (industrywide)	Cost Leadership Strategy	Differentiation Strategy
Narrow (market segment)	Focus Strategy (low cost)	Focus Strategy (differentiation)

FIGURE 2.9 Porter's generic strategies.

Differentiation Strategy

A differentiation strategy calls for the development of products that offer unique attributes and are perceived by customers as better than or different from the products of the competition. The higher value may allow the company to charge a higher price beyond the additional costs occurred in making the unique product.

To be successful in implementing a differentiation strategy a firm must have these internal strengths:

- Access to leading scientific resources.
- Highly skilled and creative product development team.
- Sales team able to communicate the perceived strengths of the product to customers.

Again, the risk is that competitors will either copy or surpass with a better product.

Focus Strategy

The focus strategy concentrates on a narrow market segment, in which it attempts to achieve either a cost advantage or differentiation. Because of the narrow market focus, firms following a focus strategy have lower volumes and thus less bargaining power with their subsuppliers. If they pursue a differentiation-focused strategy, they may be able to pass on higher costs to their customers, because no substitute products exist.

The risks of focus strategies are imitation and changes in the target markets. Other firms with a focus strategy may find new market segments they can serve even better.

After having found the right generic strategy and coming up with an understanding of the strategic advantage, it is important to make it applicable as soon as possible; it should be as great as possible, and it should last as long as possible. A strategic advantage should be ambitious, it should generate profits above the average for the industry concerned.

Hence, we always look out for potential needs for change:

- Is change necessary?
- In what direction should change take place?
- How fast will this change occur?

It is the task of *strategic management* to implement the strategy and to control the actions and the behavior required to implement change.

In technology-driven companies, two factors determine the business environment most: technology and the customers. A company has to adjust its core competencies to the needs of its customers. Depending on the business environment, we can speak of

- **Technology-driven strategy** (which is typical for startup companies).
- **Business-driven strategy** (or strategy-driven technology), which is most common for established companies. Here, the market defines the optimum strategy and choice of technology and the required competencies to execute the strategy.

Traditional mechanistic organizations typically have top-down approaches to strategy development. Planning is optimized around the original targets, and it is difficult to change directions once implementation is under way. Information about changes in the environment coming from individuals throughout the organization is not recognized.

As the organizational structure has been changing to a more organic one, strategy development started to involve the thoughts of an ever-increasing number of individuals in the organization. The strategic process is now dynamic and continuous. A change in one component can necessitate a change in the whole strategy. As such, the process must be repeated frequently to adapt the strategy to environmental changes. Adaptive processes now help companies create and adapt strategies quickly and iteratively, so that people can effectively bring up new issues and help reallocate resources in changing environments. Since people throughout the organization play roles in the company's strategic success, strategy development needs to tap into ideas from everywhere. This requires opening up the process to people throughout the organization, permitting extensive face-to-face collaboration, and arranging for individuals other than senior executives to facilitate important strategic discussions.

SUMMARY

The economic environment in which we live today is characterized, on the one hand, by worldwide overcapacity, which creates continuous pressure to reduce manufacturing costs and transfer manufacturing to lower-cost countries. In less-developed countries, on the other hand, the demand can be satisfied only by generating purchasing power.

Innovation is of ever-increasing importance in saturated markets. Major changes are due to more use of automation, information technology, and electronics. Contrary to earlier predictions, most materials generally are still available, though the increase of the world's population may cause some changes in the near future.

Knowledge is becoming a more and more scarce resource. As organizations change to more flexible ones, future careers for engineers will require a wider knowledge base.

During the course of the 20th century, major changes in how companies are organized have taken place, from mechanical, hierarchical to more organic and flexible organizations, which allow more room for frequent and rapid innovation.

The concepts of vision, mission, objectives, and strategies are used to delineate, in clear and measurable terms, where the company stands today and where it wants to be in the future. To make business strategies work, each function of a business has to develop its own part of strategy. In developing a strategy, we find the current strategic position of the company in the market and make a choice among several generic strategies as to where the future course should be directed: cost leadership strategy, differentiation strategy, and focus strategy.

REFERENCES

[1] Available at http://h30418.www3.hp.com.

[2] P. Drucker, interview by Eric Schonfeld, quoted in www.heritagetidbits.com/archives/management_and_leadership.

[3] D. Nelson, W. Frederick, Taylor and the Rise of Scientific Management, University of Wisconsin Press, Madison, 1980.

[4] R. Swedberg, Max Weber and the Idea of Economic Sociology. Princeton University Press, Princeton, NJ, 1998.

[5] H. Fayol, General and Industrial Management: Henri Fayol's Classic revised by Irwin Gray. David S. Lake Publishers, Belmont, CA, 1987.

[6] T. Burns, G.M. Stalker, Management of Innovation, Tavistock Publications, London, 1961.

[7] E.P. Learned, C.R. Christiansen, K. Andrews, W.D. Guth, Business Policy, Text and Cases, Irwin, Homewood, IL, 1969.

[8] M. Porter, Competitive Advantage: Creating and Sustaining Superior Performance, Free Press, New York, 1985.

REFERENCES

Creation Phase—Research and Development

OBJECTIVES

After reading this chapter you will be able to see research and development as an important element in the creation-application spectrum, which covers the whole range from basic research to market. With a focus on the first part, the creation phase, the chapter addresses two ways of progressing from an idea to an innovation:

1. The linear model of research and
2. The two-dimensional view of research,

and presents arguments why the wider outlook in "use-inspired basic research" is needed to be successful. The first phase of research, called the *fluid phase* or the *fuzzy front end*, is addressed by the new concept development model. In a brief overview, the various influencing factors and the five controllable activity elements are described, opening the door to a wide range of tools that can be used to narrow down the number of unknowns and speed up the process to move an idea into the development stage.

Use-inspired basic research and the many interacting factors in the early phase of research are applied to a case study on ZnO-surge arresters, where two originally separate companies tried different ways to develop the business.

IMPORTANT PROCESS STEPS FROM IDEA TO MARKET

The field of materials phenomena and techniques exploded over the past 50 years. Predictions made 10–20 years ago about the course of materials science naturally failed, because one cannot predict the unpredictable. It is therefore not surprising to see the enormous complexity that ties materials together with numerous other sciences and engineering fields. How, then, can ideas evolve just about everywhere and still find a focus in a new or improved product? Are there recipes on how to proceed, rules to be followed to secure success? The answer is yes and no with an emphasis on the *no*.

53

We have to realize that, in addition to gaining sufficient knowledge of the materials science field, many other factors play into the scenario. It is easy to grasp that each material requires a process to make and shape it, but often many processes compete with each other, so we have to first collect all possible processes and look at their limitations, with respect to the costs involved or their ability to permit full freedom of design. But, most important is the *knowledge of a need*. The need to create new knowledge is often the foundation of basic research, but the need of a customer is what leads to a new or improved product that motivates us to apply knowledge purposefully.

Looking at the complexity in materials science and identifying a need is still only one side of the coin. *People* are involved along the whole chain of events. Typically, a material scientist coming straight from university is still focused on finding original research ideas leading to new knowledge, some of which he or she may see published. The scientist eventually learns to be *open and knowledgeable enough to branch out into totally new areas*. As a young powder metallurgist, a couple of years after leaving the field of basic research in powder metallurgy at the Max-Planck-Institute and starting to work in an industrial research laboratory, I remember making this comment: "I didn't know the difference between superalloys and superplasticity (meaning I didn't know anything about either), but we are now among the world leaders in research in these two areas, because we have internal customers who came to us with a need."

Sometimes, but very rarely, one person or one group of people drives this process through from beginning to the successful end. More likely, a combination of many events will bring together the required set of skills, often by coincidence. When a larger group of dispersed people is involved, the question arises how to bring them together or connect them through a suitable organization and how to best manage the process from idea creation to marketing a new product.

If we leave out the persons involved and where they were for a particular activity and look only at the logical sequence of events as a function of time, we come up with a one-dimensional structure. Several terms have been used to describe it: *linear model*, *fuzzy front end*, *the new product development process*, and *commercialization*.

Over time, these models have evolved through four different phases [1]. The first two involved linear flows of knowledge. From its beginnings in the mid-19th century to the 1950s, the process was based on serendipity and somehow isolated from the other functions of the firm. In the 1950s and 1960s, it adopted the basic routines of project management. In the 1970s and early 1980s, project and business development groups appeared within the firm to coordinate different functions and assure a multidirectional flow of information. In the late 1980s and 1990s, technological alliances with users, suppliers, and competitors increased the nonlinear flows by incorporating information generated outside the firm.

Before entering the discussion on the flow of innovation, let us start with the process steps. We can sum up all the individual process steps in the *creation-application spectrum* (Figure 3.1). Looking backward, we can identify the

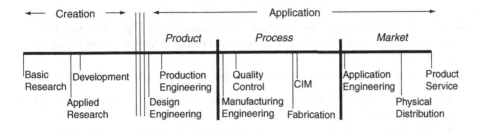

FIGURE 3.1 Creation–application spectrum.

individual process steps. The challenge arises when looking forward, since there are many unknowns. We therefore have to look in more detail into the different aspects of the whole process:

- Research and development in the creation phase.
- Design, processing, production, marketing, and sales in the application phase.
- Economical considerations in funding research.
- Finding the most cost-effective production technologies.
- Advantages and disadvantages of various organizational concepts.
- Finding the best way to manage people and the whole process.

The literature is filled with tools to speed up the processes, but there is one simple reasoning. To use a tool effectively requires the precise knowledge of the right processes and all the elements of it. Very rarely do we know all the process steps in great detail, unless we deal with incremental improvements in an otherwise well-established process.

CREATION PHASE—RESEARCH AND DEVELOPMENT

The process starts with the creation phase, consisting of research and development. What is the ideal scientist doing? Field [2] describes it as the cycle of proof, the scientific method. The scientist starts out with gathering all available data into a *database*, by reading scientific literature and attending conferences or meetings with scientists working in the same field. Then, looking at all the available data, the researcher comes up with an assumption, a *hypothesis*, of how the different ideas correlate. From the hypothesis, the scientist makes a *prediction* of the outcome of an *experiment*, which he or she and perhaps members of a whole group will be able to do, assuming they have the expertise and the resources. The purpose of the experiment is to *verify* if something is true or not. More often than not, a prediction turns out to be wrong. The scientist then goes back to the starting point and reviews how the experiment was set up and carried out. The scientist may have to modify the hypothesis or come up with a totally new one. The work described here, in this cycle, is usually called *basic research*, simply defined as the work that helps increase scientific understanding to expand human knowledge. There is no obvious commercial value yet to the discoveries that result from basic research.

For example, basic science investigations probe for answers to such questions as

- How do you predict phase diagrams from thermodynamic data of the starting components?
- How do you explain the strength of existing materials and predict the strength of new ones?

Applied research is designed to increase practically usable knowledge or solve practical problems of the modern world, rather than acquire knowledge for knowledge's sake. For example, applied researchers may investigate ways to

- Find new methods to produce biomedical components.
- Improve the energy efficiency of homes, types of power generation, or modes of transportation.

Development is the concretization of research results with respect to an economic application. For example,

- Showing the feasibility of producing net-shape titanium alloy components by isothermal forging.
- Demonstrating a low-cost manufacturing technique for small stationary fuel cells.

Product development starts with a specification for a new product or a new process. A list of specifications includes technical ones, target markets, and target sales volumes. Examples would be

- The use of carbon-nanotube-reinforced epoxy in sports equipment for the upper-end market.
- The use of superelastic shape-memory alloys in eyeglass frames.

There are several main objectives of R&D, depending on which angle of view we take:

- Because of the early recognition of new technical solutions and their opportunities and risks, R&D is always an important contributing factor, when *formulating goals and strategies of a company*.
- R&D has to *provide technical solutions needed to reach a target*, such as prototypes and manufacturing processes, with optimum use of timing and funding.
- To translate technical knowledge into economic success, it has to go beyond technical feasibility and demonstration to *demonstrate economic success*.
- It can make an important contribution to *changing the culture in a company* as a result of changing market requirements.

An R&D project, to be successful, has to meet technical specifications and economic goals, which are to maintain or increase the economic success of a company by

- Continuous improvement of existing products.
- Development of new products.
- Substitution of outdated products.
- Production improvements (rationalization, lowering of quality costs).
- Improvements in procurement (other materials or suppliers; make or buy decision).
- Diversification (new products for new markets).

As mentioned earlier, if we look at only the timescale for the development of a new product, we miss many *sources of technical knowledge* available to a company. Size and maturity play an important role, but still many options exist. Especially with large companies, *internal R&D* is the most common way of generating and using new knowledge. To reduce the high-risk burden of basic research, companies cooperate with *external research institutes* or *universities*. Anticipating future cooperative supply networks, a company may decide to cooperate with a *supplier*. A fairly recent example of that was the introduction of single crystal superalloy turbine blades into industrial gas turbines. Working with an established precision casting company not only allowed the user company to save on expensive investments in research equipment, it also helped to speed the transfer of the technology into the production stage. Finally, as a frequent way to quickly absorb the results of R&D, companies may decide to forego internal R&D and buy the technology developed by a *third company*.

R&D has always been considered a cost factor, an investment into the future. Funding can come from many sources:

- *Government funding* is often made available, especially in the early, high-risk stages of a project.
- *Indirect government funding* becomes available when a company decides to cooperate with a government-funded organization.
- In later stages of projects, *mixed external and internal funding* is a well-established route in many countries.
- A very successful way, however, is to switch to *internal funding*. This can be on a corporate level in large companies or in business units responsible for either existing or new products that can fund R&D projects directly.

The Linear Model of Innovation

The linear model is a simplified model of how innovation occurs. Reality is more complex. There are stronger feedback loops among stages (Figure 3.2). Research may often be market or need driven. Research is relevant to all stages. Multiple actors press on their priorities (e.g., government, industry).

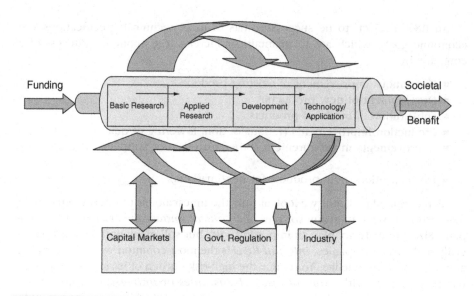

FIGURE 3.2 The linear model of R&D leading to innovation.

Looking at Figure 3.2, it does not seem to be illogical to assume that the shortest way from a research idea to a marketable product is to follow the linear path. This has long been considered the winning strategy.

The American research agenda after World War II was greatly influenced by Vannevar Bush's famous report to President Roosevelt [3], in which he writes: "Basic research leads to new knowledge. It provides scientific capital. It creates the fund from which the practical application of knowledge must be drawn. New products and new processes do not appear full-grown. They are founded on new principles and new conceptions, which in turn are painstakingly developed by research in the purest realms of science.... Today, it is truer than ever that basic research is the pacemaker of technological progress."

Since then, he has been, perhaps incorrectly, called the *creator* of what has been characterized as the well-known "linear model" of innovation, described earlier, indicating a continuous flow from basic to applied research and from applied development to technology transfer to practical applications. Basic research was considered the fuel that drove the enterprise.

The model also assumes that government money is required to fund academic science, but many further propose that governments should also fund technology development (Figure 3.3).

Data have shown that only 10% of new technology emerges from academic research. Ninety percent of new technology arises from the industrial development of preexisting technology, not from academic science. Both science and technology are largely self-contained, growing on themselves. Basic science is based mostly on old science; technology development is based mostly on old technology.

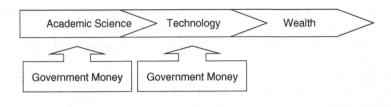

FIGURE 3.3 Funding in the linear model of innovation.

In many countries, it is still strongly believed that government funding can help develop new technology from the old one. There are several reasons why this assumption often may lead to failure. Governments can raise money only through taxing industry. Because the money comes from taxes, there often is an inherent drive to create new business and new jobs by funding research topics, which on a government level are considered most promising. To fund and direct its R&D, a company would suppose that government knows better than the directors of the enterprise how best to optimize that enterprise's R&D.

An important aspect is to decide when to engage in a new research area. Kealey [4] explained how such decisions lead to first and second mover advantages.

First Mover Advantages

These advantages are assumed to be generated from discovering something first. We can also say that this comes from *technology-push R&D*. The commercial fruits can be huge, but basic research is unpredictable. Normally, it is too risky for generous industrial support. But, to be successful, high-risk projects also have to be carried in the portfolio of research projects of larger companies, and often startup companies risk their whole future on such projects. An example of a first mover project is the development of superconducting fault current limiters.

Second Mover Advantages

The process is something like jumping on a moving train. Copying is much easier than doing original research, and the second prize can be bigger than the first. Companies pay scientists to perform first-mover research, but they expect to gain more benefit from second-mover research (work done in the library). Also, one country has to fund first-mover domestic R&D to generate the competence to capture second-mover R&D.

The Real Economics of Research in the Linear Model

Experience also shows that government and private funding usually follow simple rules:

- The richer the country, the more research it does.
- State funding of research displaces private funding.
- Funding is disproportionate, since the state displaces more private funding than it provides.

Very often top management is driven by accountants, who argue that the company can save funds for R&D investment by using government funds—and it can save even more if it does not spend an equivalent amount on its share of its contribution.

The linear model of research was quite common in the United States. Large companies like IBM, GE, and Bell, set up large basic research centers, which recruited the best scientists of the time, who were more or less free in their ideas to pursue basic research. To some degree this model was copied by large European companies in the late 1960s and early 1970s. An example is the Brown Boveri Corporate Research Center in Switzerland in 1967.

In the words of A.P. Speiser, these were important points of consideration when establishing the research center [5]:

Corporate Research was founded as a result of a need that developed during the 1960's. During the postwar years, scientific knowledge, particularly in the area of physics, increased tremendously. Parallel to this development was an increasing trend towards specialization. With the foundation of Corporate Research at Brown Boveri contact with the outside scientific community was increased and at the same time a contribution was made to the general pool of knowledge from which one may not draw without making some contribution in return. One of the important tasks of Corporate Research is to carefully observe advances made in the outside world which could lead to possible applications and to meet the challenge of eventually transforming scientific knowledge into a technologically useful form.

Not surprisingly, establishing a top-down research center within a company that already had many well-established test centers, even central laboratories, was met with much disbelief and resistance from the established organization. The research center met its first goal quickly, to recruit first-class researchers and gradually establish an organization according to scientific disciplines, each of which could simultaneously influence several product areas. It also established a portfolio of medium- to long-term research projects. At least two projects can be named as examples of following the linear model of research. One was the development of a new class of permanent magnetic materials with the vision to use them in the design of permanent magnet motors; the other was the development of liquid-crystal displays in cooperation with the chemical company Hoffmann-La Roche and later in a joint venture with Philips, called Videlec. Both developments, although successful technically, did not become commercial successes in Brown Boveri; but, on a worldwide scale both developments resulted in huge businesses outside the founding company. The LCD business moved to the Far East, and Brown Boveri (later ABB) could at least cash in on license fees on the order of several hundred million dollars. Permanent magnet motors also became an established business within ABB, where they are used as motors attached to the rudder of a boat, driving ship propellers, thus allowing large ships to turn around in a small radius. Due to the large time difference of in-house development of permanent magnets and their application in motors (about 30 years), the material could be purchased from outside vendors. For a long period, the business units upheld a negative attitude toward the "arrogance" of the scientists. Breaking down these barriers and coming to cooperative agreement between the business units and the existing test laboratories remained a slow, tedious process for a long time. It could be resolved only by moving researchers into business units or bringing future project leaders from the business unit into the research center to make them familiar with the results of some of the research projects.

Two-Dimensional View of R&D

The creation of the Brown Boveri Research Center is not an exact example of the linear model of innovation. The dream has always been to come up with ideas that would lead to products and new business. Therefore, the one-dimensional viewpoint, which has been so influential in America, can now be considered too narrow. Stokes [6] convincingly describes a broader view of how to think about research efforts. Stokes suggests that a two-dimensional view (Figure 3.4) is more descriptive of what really happens than the one-dimensional "basic to applied" research concept. His two-dimensional view is that *pure* research, such as the work of Niels Bohr on atomic structure, can be thought of as a vertical axis of research space, while the strictly *applied* research of Thomas Edison can be visualized as a horizontal axis. Stokes calls the area between the axes *Pasteur's quadrant*, because the work of Pasteur exemplifies clearly the interwoven nature (not one dimensional) of basic and applied research. The concept is that research like Pasteur's often is applied, practical, and basic at the same time. He calls it *use-inspired basic research*. It seems likely that research in the coming years will be

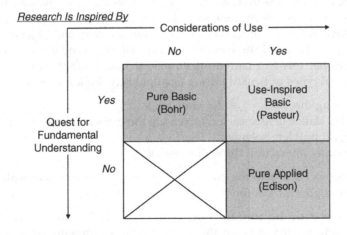

FIGURE 3.4 Stokes's R&D matrix and Pasteur's quadrant.

considered superior if it produces practical results and promotes basic understanding. A wider scope of results than may have been expected in the past is likely to be desired from those participating in future R&D efforts. Work will be expected to occur well within the horizontal and vertical borders of Pasteur's quadrant.

Some words of caution should be added here. While the idea of use-inspired basic research sounds convincing, we have to realize that there will always be continuous fluctuation among all the quadrants. Also, not all researchers are the same. Some have greater inclination toward basic research; others might prefer to be engaged in applied research. It is the task of the leadership to balance the various interests and talents in the direction leading to the best and fastest development of new products.

The case study on surge arresters (varistors) later serves as an example of use-inspired basic research.

Benefits of Academic Research and Government Funding

Undoubtedly, economic benefits arise from academic research and public funding [7,8], although a direct correlation, like rate of return for the investment by the government, is hard to prove. Let us begin with an example of personal experience to illustrate the complexity. Starting out with a thesis in powder metallurgy supported by the U.S. Atomic Energy Commission and working in R&D in this field for about four or five years in Germany and Switzerland, the work soon moved in the direction of manufacturing of Ti-impeller wheels. Very quickly, the approach changed from powder metallurgy to superplastic deformation. The conclusion was this: The skills developed in the United States and especially the acquired skills in powder metallurgy led to a contact from the development department of a business unit, which very soon was directed toward a totally different technology. Definitely, new skills became necessary to carry out the project, but there was no connection between the location and objectives of the funding organization and the content and location of the new research project. Considering that this is a fairly normal way of how innovation is created, we can list several types of benefit arising from publicly funded research:

- Increasing the stock of useful knowledge.
- Training skilled graduates and increasing the capacity for scientific and technological problem solving.
- Stimulating interaction through networks.

Other factors come into the picture if one wants to address economic benefits coming out of government-funded projects or from university projects:

- **Top-down or bottom-up planning**. Experience shows that the latter approach, leaving it up to the researcher at a university or in a research organization to promote his or her idea and search for the best partners, always works much better than the top-down approach. Here, bureaucracy may come up with unrealistic requirements: Having partners from several countries (in the European Union) or from other universities in one country often leads to groups where some of the partners make no contribution and administrative costs become a sizable part of the total budget.

- **Use-inspired basic research versus pure basic research**. A research project leads much faster to commercial success if a researcher is driven by the vision of applications in addition to an interest in creating new knowledge.
- **A "pull" from the side of the user of the results**. In many cases, industrial companies are better able to recognize new needs from the market. If they have close personal interaction with an academic institute, they can help define the right direction for the research project.

Three Phases of the Innovation Process by Abernathy and Utterback

Abernathy and Utterback [9] describe the innovation process in three phases: the fluid phase, the transitional phase, and the specific phase.

Fluid Phase

In the first phase, technological and market uncertainties prevail, a great deal of change takes place simultaneously, and outcomes may vary significantly. There is almost no process innovation, and the many small firms competing base their advantage on differentiated product features. Competition is not as stiff, because companies have no clear idea of potential applications for the innovation or in what direction the market might grow. The major threats come from the old technology itself and the entrance of new competitors if the innovation is radical.

In this phase, a company can either establish its product as a *dominant design* or prepare itself, in a waiting position, for the appearance of the dominating design. Then, once the standard becomes clear, it will try to secure most of the profits, basing its competitive advantage on the distribution channels, supplier contracts, complementary technologies, the like: "The *dominant design* product has features that competitors and innovators must adhere to if they hope to command significant market share following" [10].

Transitional Phase

As producers start to learn more about the technology application and customer needs, some standardization emerges. Usually, by this time, the acceptance of the innovation starts to increase and the market starts growing. The convergence pattern in this phase leads to the appearance of a *dominant design*, which is a product design whose main components and underlying characteristics do not vary from one model to another.

Winning the battle for the dominant design enables the firm to collect monopoly profits, assuming that it cannot be copied easily or intellectual property rights are secured. Even if the standard is "open," the developer can build complementary products or enhanced versions faster, possibly establishing a new standard in the future. Firms in this phase use strategies to consolidate their product positioning and start increasing production capacity and process innovation to face the next phase, the specific phase.

Specific Phase

After the appearance of the dominant design, competition shifts from differentiation to product performance and costs. Companies now have a clear picture of market segments and therefore concentrate on serving specific customers. Manufacturing uses highly specialized equipment and can start employing less-skilled labor for higher-volume production.

Conventional and Disruptive Technology S Curves

The S curve illustrates the introduction, growth, and maturation of innovations as well as the technological cycles that most industries experience. In the early stages, large amounts of money, effort, and other resources are expended on the new technology but small performance improvements are observed. Then, as the knowledge about the technology accumulates, progress becomes more rapid. As soon as major technical obstacles are overcome and the innovation reaches a certain adoption level, an exponential growth takes place. During this phase, relatively small increments of effort and resources result in large performance gains. Finally, as the technology starts to approach its physical limit, pushing the performance further becomes increasingly difficult.

Christensen [11] emphasizes the importance of technology S curves in technology management. For a given product in an established market, the main focus of technology management is to look out for the next generation of technology and introduce it at the right time (Figure 3.5). A good example is the first introduction of precision-cast cooled gas turbine blades (technology 1) followed by uncooled single-crystal superalloys (technology 2), which then were further upgraded by the application of TBC (thermal barrier coatings) on cooled single-crystal superalloys (technology 3). Each new technology was allowed to further increase the high-temperature capability of the blade material, allowing an increase in the thermal efficiency of the gas turbine.

A *disruptive technology* emerges and progresses independent of an established one, therefore its performance may have to be measured differently from the existing one. If and when it progresses to the point that it can satisfy the level of performance demanded by the established technology, the disruptive technology can then invade, replacing the established technology. Figure 3.6 shows two

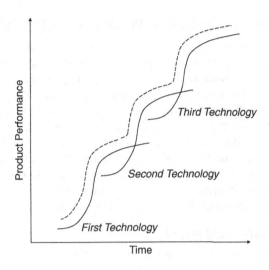

FIGURE 3.5 The conventional technology S curve.

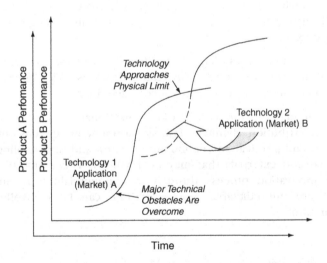

FIGURE 3.6 Disruptive technology S curve.

superimposed plots, the existing and the disruptive innovation. Two different y axes have to be chosen, because different performance parameters may apply. The introduction of the hybrid car is a good example of a disruptive technology; superconducting fault current limiters and fuel cells, if ever successful, would be other examples.

Models and Tools for the Start of the Innovation Process

As described in Chapter 1, we have to introduce additional dimensions to describe the innovation process. We already discussed the many possible sources of innovation. We now want to look at the beginning of the process of innovation. Koen et al. [12] present a good overview of the topic. They divide the innovation process into three areas: the fuzzy front end (FFE), the new product development (NPD) process, and commercialization, as indicated in Figure 3.7.

The figure shows that the beginning phase is the least well defined, some thinking was applied to find the influencing factors that lead from a new opportunity or new idea to a concept. The new concept development (NCD) model [13] evolved from an analysis of many case studies and lists all possible factors to be considered to optimize the front end of innovation.

New Concept Development Model

The NCD model consists of three parts (Figure 3.8):

1. The *engine portion* is the leadership, characterized by early involvement of a business executive champion, setting aggressive targets, and maintaining a culture that encourages innovation and creativity. The entire innovation process (including both FFE and NPD) needs to be aligned with business strategy to ensure a pipeline of new products and processes with value to the corporation.

2. The inner spoke area defines the *five controllable activity elements*: opportunity identification, opportunity analysis, idea generation and enrichment, idea selection, and concept definition.

3. The *influencing factors* consist of organizational capabilities, the outside world (distribution channels, law, government policy, customers, competitors, and political and economic climate), and the enabling sciences (internal and external) that may be involved. These factors affect the entire innovation process through to commercialization. Since many events are unpredictable, not everything can be controlled by the corporation.

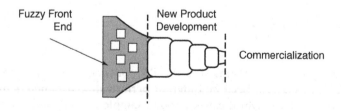

FIGURE 3.7 Three areas of innovation process [9].

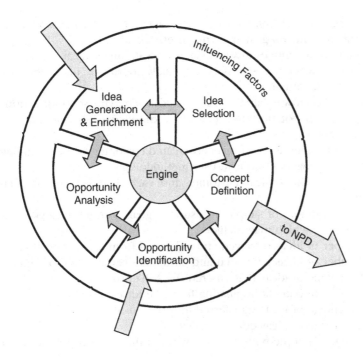

FIGURE 3.8 New concept development model (Koen et al. [13]).

The five controllable activity elements contain a list of all possible actions and refer to existing tools [13].

In **opportunity identification**, the goal is to create more opportunities by envisioning the future. This can be accomplished by

- Roadmapping.
- Analyzing customer and technology trends.
- Competitive intelligence analysis.
- Market analysis.
- Scenario planning.

An example of the last point is the request from GE's former CEO Jack Welch: "Destroy your business." Each business unit had to prepare a plan of how, as competitors, it could erode GE's customer base. Envisioning a new future through the eyes of competition triggered a new strategy of how to deliver appliances directly from its own warehouses, thereby strongly reducing the transaction costs.

Idea generation and enrichment covers the birth, development, and maturation of a new idea. The whole process is evolutionary, as many iterations and changes are very likely. Some of the techniques include

- An organizational culture that encourages employees to spend free time testing and validating their own and others' ideas.
- A variety of incentives (awards, prizes) to stimulate ideas.
- A process owner, whose task is to formally coordinate ideas from generation through assessment.
- A mechanism how to handle ideas outside or across the scope of a business unit.
- Frequent job rotation to encourage knowledge sharing and extensive networking.
- Methods for identifying not yet articulated customer needs, including ethnographic approaches and lead user methodology.
- Discovering the typical customer and early involvement of a customer champion.
- A limited number of simple, measurable goals to track idea generation and enrichment, which may include
  - Number of ideas retrieved and enhanced from an idea portfolio.
  - Number of ideas generated and enriched over a period of time.
  - Percentage of ideas that entered the NPD process.
  - Number of patents originating from the ideas.
  - Percentage of ideas transferred into the business units.
  - Percentage of ideas commercialized.
- Inclusion of people with different ways of thinking on the idea enrichment team.

Idea Selection

Very often, there is no shortage of new ideas. The problem for most businesses is to select the right idea, that with the highest business impact. Formalized decision processes in the FFE are difficult, due to the lack of information and understanding available at such an early stage. Although various forms of business impact calculations are used very early on, they are often wild guesses at this stage. Idea selection often begins as individual judgment, which may occur subconsciously. As several of the case studies in this book show, often early personal judgments are made at an intuitive or "commonsense" level, with little more than the idea itself to consider. Idea selection within an individual's mind are almost always the initial part of the selection process.

In spite of these early difficulties, a formal process for idea selection has proven to be useful and necessary, otherwise new ideas might disappear and be forgotten. Continuous feedback and communication with the originators of ideas is important to keep up their motivation to produce new ones. Some of the more effective tools and techniques deal with

- Portfolio methodologies based on multiple factors (not just financial justification):
  - Technical success probability.
  - Commercial success probability.

       □ Reward.
       □ Strategic fit.
       □ Strategic leverage.
- Formal idea selection processes with prompt feedback to the idea submitters:
  - □ Enhancement of methodology with electronic performance support systems.
  - □ Web enabling of the process (see also "Stage-Gate® Process" in Chapter 6).
- Use of options theory to evaluate projects.

Concept Definition

Concept definition is the final element of the NCD model. If successful, the project can enter the NPD stage, where serious investment decisions will be made. Typical guidelines address

- Objectives.
- Fit of the concept with corporate or divisional strategies.
- Size of opportunity, such as financial impact.
- Market or customer needs and benefits.
- A business plan that specifies a specific win/win value proposition for value chain participants.
- Commercial and technical risk factors.
- A project plan including resources and timing.

Some of the processes for concept definition are

- Setting criteria for the corporation that describe what an attractive (in terms of financials, market growth, market size, etc.) project looks like.
- Rapid evaluation of high-potential innovations.
- Rigorous use of the technology Stage-Gate process for high-risk projects (see Chapter 6).
- Understanding and determining the performance capability limit of the technology.
- Early involvement of the customer in real product tests.
  - □ Involvement of the customer even before product is completed.
  - □ Staff up high-potential projects while still in FFE.
- Finding partners outside of areas of core competence.
- Focus.
- Evaluating alternative scientific/technical approaches.

To sum it up, ideas and opportunities can evolve from just anywhere, from marketing in *market pull* and from R&D in *technology push*. There are many potential triggers for an innovation. Therefore, there is no predefined order in which one has to work; it is rather a continuous interaction between these activities. Trott [14] calls this the *interactive model of innovation* (Figure 3.9).

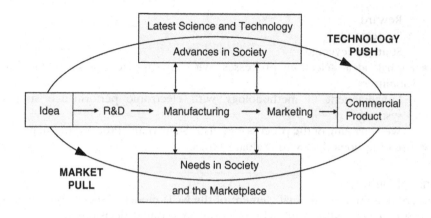

FIGURE 3.9 Interactive model of innovation (Trott [14]).

Table 3.1 Difference Between the Fuzzy Front End and the New Product

	Fuzzy Front End	**New Product Development**
Nature of work	Experimental, often chaotic. Can schedule work—but not invention	Disciplined and goal-oriented with a project plan
Commercialization	Unpredictable or uncertain date	High degree of certainty
Funding	Variable—in the beginning phases many projects may be "bootlegged," while others need funding to proceed.	Budgeted
Revenue expectations	Often uncertain, with a great deal of speculation	Predictable, with increasing certainty, analysis, and documentation as the product release date gets closer
Activity	Individuals and team conduct research to minimize risk and optimize potential	Multifunction product or process development team
Measures of progress	Strengthened concepts	Milestone achievement

Table 3.1 shows in simple terminology the differences between the fuzzy front end and the new development process. The simple lesson from looking at the table is that there will never be cookbook recipes how to manage the FFE part. This is clearly demonstrated by a few of the case studies, which show that

a clear goal at the beginning of the idea creation process may have to be replaced by a new goal because of unexpected obstacles or new discoveries along the way.

New Process Development [12]

Although all kinds of approaches always have been used to come up with an innovation, historically, there has been a change of emphasis, coupled with changes in organizational thinking (modified from [13]):

1960s–1970s: Technology push (following models of IBM, GE).
1970s–1980s: Market pull.
1980s–1990s: Combination of push and pull (interactive model).

Case Study. Metal Oxide Varistors, Evolution of a Technology: In-house Research or Licensing? (C. Schüler [15])

Surge Arresters

Under normal operating conditions, electrical transmission and distribution equipment is subject to voltages within a fairly narrow range. Due to lightning strikes, switching surges, or other system disturbances, portions of the electric system may experience momentary voltage levels that greatly exceed the levels experienced by the equipment during normal operating conditions. Left unprotected, costly equipment, such as transformers, switching apparatus, and electrical machinery, may be damaged or destroyed by such overvoltages and the resultant current surges. Accordingly, it is routine practice within the electrical industry to protect such apparatus from dangerous overvoltages through the use of surge arresters.

A surge arrester is commonly connected in parallel with a comparatively expensive piece of electrical equipment, to shunt or divert the overvoltage-induced current surges safely around the equipment, thereby protecting the equipment and its internal circuitry from damage. When caused to operate, a surge arrester forms a current path to ground having very low impedance relative to the impedance of the equipment it protects. Once the transient condition has passed, the arrestor must open the recently formed current path to ground and again isolate the distribution or transmission

The previously used conventional arrester is commonly referred to as the *series gapped silicon carbide* (SiC) *arrester*. The nonlinear resistive elements in this arrester are relatively short cylindrical blocks of silicon carbide stacked one atop the other within the arrester housing in series with spark gap assemblies that are generally resistance-graded gap assemblies. A resistance-graded gap assembly consists of a resistor electrically in parallel with the spark gap and usually includes one or more resistors in series with the gap. This network of resistors is employed to control the voltage level at which the spark gap will begin to conduct.

In the 1980s, a new, superior surge arrester concept replaced the previous technology, which had been successfully used over the past few decades. New voltage-dependent resistors were developed, so-called metal oxide varistors (MOVs), which no longer required dischargers. In addition, they provided better protection than the old technology.

Because of the different degrees of nonlinearity of the resistive elements employed in silicon carbide and MOV arresters, these arresters differ in structure and operation. The silicon carbide blocks are designed to provide a very low resistance to surge currents but a higher resistance to the 50–60-Hz power-follow current, which continues to flow through the arrester after the transient condition has passed. Despite the higher resistance, the silicon carbide blocks still conduct large currents at the normal, steady-state, line-to-ground voltage. Accordingly, gap assemblies are employed in series with the silicon carbide blocks. As a transient overvoltage condition ceases, the resistance of the silicon carbide blocks increases, to limit the magnitude of the power-follow current. The reduced current flow and the corresponding decrease in the voltage across the spark gaps provide the gap assemblies the opportunity to open the current path to ground and thus "reseal" the power circuit after the surge has passed.

The new technology was successfully introduced in high- and medium-voltage applications. Both Brown Boveri and Asea, the predecessors who later merged to form ABB, had used different approaches to develop the technology and invest in manufacturing, bringing successful products to the market.

A critical review of the various paths taken by the two companies shows that there is no one way but always several possible ones to become successful. Still, the questions arise, Which approach was better, and can we learn anything about future development projects? Also, which obstacles had to be overcome to make the technology a success?

The History of ZnO Varistors

The first work on nonlinear electrical properties had been initiated in the former Soviet Union, but the potential of the application of the effect in products was ignored and forgotten. This was a not atypical result of basic research in that country at the time. There were excellent research centers in many places in the country, but when asked about potential applications, the researchers often replied, "Our task is to conduct basic research, not to develop applications—this is the task of some other institute."

The material was rediscovered by Michio Matsuoka at the Wireless Research Laboratory of Matsushita Electric Industries in Japan in 1968. The first research focused on surface varistors, but later, the bulk effect was developed, and efforts to develop a product were initiated. In 1972, the material became available in the United States, where GE had taken a license from Matsushita, under the trade name GE-MOV varistor, mainly for the protection of consumer and industrial equipment, working below voltage surges of 1000 V.

Metal Oxide Varistors at Brown Boveri

Brown Boveri, through its Swiss Corporate Research Center in Dättwil/Baden learned about metal oxide varistors when GE announced its first product in a press release in 1972. Via the U.S. division of Brown Boveri, the researchers in Switzerland quickly got more detailed information about the new material. The components were tiny, the size of a coin, but the electrical properties were fascinating, especially the exceptionally strong voltage dependence of the resistance. It took little imagination to think of the innovation potential of the material if used in high-voltage surge arresters, provided one could develop components of the required size and power range.

This was exactly the vision the researchers were ready to embrace. They had just scrapped a project directed to improve the properties of the established spark-gapped silicon carbide resistors, called *RESORBIT*. The work leading to a better basic understanding of the material had surprisingly failed to lead to technically usable results. In spite of these negative results, permission was given to start a new project, metal oxide varistors, in 1974, which even was partially supported by the business unit, although those in the unit had their doubts whether upscaling from the present low-voltage "dwarfs" to high-voltage "giants" could succeed and replace the well-established RESORBIT, which was based on silicon carbide, sintered together following secret recipes.

About two years later, in mid-1976, the research center obtained promising results with blocks of varistors for the kilovolt range. Even more important was the newly found understanding, how voltage-dependent resistance (varistor effect) is physically related to the special microstructure obtained after sintering the ceramics. The promising results came just in time, because a flood of publications and patents indicated how much GE and mainly Japanese companies (Matsushita, Toshiba, and others) were investing in the development of high-voltage varistors. Every conceivable variant of material composition and processing technique was tried to improve the electrical properties.

The results of the researchers in Brown Boveri Corporate Research and other research groups made it clear, around 1976, that the new metal oxide varistors would soon replace the existent surge arrester technology. Many potential manufacturers started to transfer the work from the research center to more realistic testing conditions in high-voltage labs. There was a continuous flow of information about progress and remaining problems with the new components, both in technical journals and at technical conferences.

Not surprisingly, the optimism of the researchers was not shared by the surge arrester business unit: The remaining risks were considered to be too high, the manufacturing costs seemed to be too high compared with the costs of RESORBIT, and uncertainty remained about the untested lifetime of the metal oxide components. The technical director of the business unit still believed in the potential of a competing technology, a next generation of the magnetically blown spark gap. The business manager questioned a foreseeable rate of return from a comparatively small business requiring a very large investment. He went further and asked for a return from a not yet completely developed product, before making his own major investment. It took the personal intervention of the highly reputed research director to redirect the project, "research is no substitute for development." The most that could be accepted by the product engineers were special applications with low market volume. In 1980, a pilot line for manufacturing was installed, but this was driven more as an insurance against failure than by a need to succeed in the market. Also, a cultural gap remained between the scientists in the research center and the production engineers in the newly established pilot plant. The former were still driven primarily by interesting scientific questions, the latter needed help in getting practical solutions.

Another drawback in the early part of the project was the rigid line organization, which prevented target-oriented development. The functions of sales, engineering, and production were distributed in special separate units; purchasing and order processing were part of financial management; R&D was again a separate unit. The structure was not organized to follow the process from development to market, and marketing and sales capacity were lacking.

Asea Bets on Metal Oxide Varistors

Asea, at the same time, moved much faster. Although no project done at its research center prepared the business unit for the new technology, Asea had more forward-looking development engineers. The main drivers for innovation were in the business units. People there had carefully observed the evolution of the new technology in the outside world. As the flood of information continued to rise, they came to believe in the future of metal oxide varistors as the new surge arrester technology. In a less hierarchical organization, decision paths were shorter at Asea. In 1976, the business area invested in a wide license from Matsushita to use that company's patents and manufacturing know-how. Without further delay, manufacturing equipment was built and installed and initial manufacturing problems were quickly overcome. While Brown Boveri was still in the evaluation phase, Asea's production was already onstream.

Brown Boveri Closes the Gap

In spring 1980, an Asea flyer caused strong irritation among Brown Boveri's management. Asea was offering a complete range of high-voltage surge arresters made of metal oxide varistors and no more dischargers. At the same time, Brown Boveri was still conducting the first field tests with surge arresters from the pilot plant. No decision had been made about commercial production. Of course, the decision was now made immediately.

A race had started to close the gap as quickly as possible and participate without delay in the beginning business with the new technology. Fortunately for Brown Boveri, other big competitors entered the field, among them General Electric, who were willing to even supply varistors to Brown Boveri because of their initial overcapacity. So Brown Boveri could pursue orders before its own production was running; the sales division could keep its position with its established customers.

After management in Brown Boveri decided to switch to the new technology, the demand for support from specialists in the research center was revived. A new research project was requested and started within the context of corporate research.

From the beginning, Brown Boveri developed its own technology to produce varistors. The chemical composition of the powder mixture and the right processing steps for pressing, sintering, passivation with glass film, and contacting were the result of a careful optimization process, at first in the research center then in the pilot production. Based on these results, very close cooperation between corporate research and development and manufacturing in the business unit was established.

The technology transfer was sped up by transferring the key project leader from corporate research to the business unit and making him responsible for varistor development. All the knowledge about the material, its processing, and the physical understanding of electrical properties, resulting from previous research, were now fully available to the business unit. In 1985, a new director of engineering was hired. Soon, he took over the functions of marketing, sales, and production. The close subsequent cooperation between research, product development, and production was the main reason, in 1982–1987, why Brown Boveri could make good on its lag and regain its internationally strong position.

The researchers also succeeded in formulating theoretical models explaining the physics of the varistor effect and proving them quantitatively through experiments. Their research became part of the classical base in this field and was responsible for ABB metal oxide surge arresters becoming internationally renowned.

Competitors Become Partners

The merger of Brown Boveri and Asea in 1988 helped make the combination a leader in the worldwide surge arrester business. This position was further strengthened by the acquisition of Westinghouse's surge arrester business. At that time, Asea's annual turnover of varistors in Ludvika was about twice that of Brown Boveri's in Baden, Switzerland. This marked advantage was due to the courageous decision, in 1976, to aggressively invest in the new technology and, as quickly as possible, offer a complete range of products, covering the whole market. The strategy was a success, although a number of technical obstacles had to be overcome. Due to its market leadership, the leadership role in the business unit went to Ludvika in Sweden, but manufacturing in Switzerland was kept alive, due to good technical quality of the products and an established customer base. Research continued, now providing results to both manufacturing sites.

A few years later, most of the high-voltage varistor production was concentrated in Ludvika, but Baden faced the challenging task of developing a new portfolio of medium-voltage varistors, which was accomplished after a few years of intensive cooperation between research, development, and production. Still, Baden also produced some high-voltage surge arrester for gas-insulated and air-insulated switchgear.

Learning Points

Brown Boveri corporate research, although very strong in basic materials research, had an early understanding of the potential of metal oxide varistors for high- and medium-voltage varistors. It failed, however, to close the "credibility gap" with the management of the surge arrester business. The researchers lacked the mindset of electrical engineers who develop devices for electric power transmission in close cooperation with the customer. The know-how base in the research center, on the other hand, was in the hands of physicists and materials scientists. They were able to develop good varistors but unable to find the right language to argue convincingly for the superiority of the varistors in practical surge arrester technology. After some delay and because of visible progress by the competition, they succeeded with their arguments, but by then, it was too late and a good part of the original technical advantage had been lost.

The credibility gap was enhanced by the earlier competition between the business unit and corporate research about which is the right new technology, which probably helped raise doubts in higher management about the usefulness of investing in the new technology.

It is only natural that research projects with a large innovation potential have to overcome big hurdles when transferring the results to the business unit. Much can be gained or lost. It is important that project management recognizes early enough that the project group has the same understanding of the application technology as the business unit. Unless this task is successfully mastered, failure is very probable. Responsibility for project management should be transferred to the business unit as early as possible, by transferring a key researcher to the business unit, training a development engineer from the business unit in the research center or as is most often done, by involving people from all the organizational units in a team effort.

This case study certainly shows how work was done in use-inspired basic research and points to unexpected obstacles across functional units and the need to break down communication barriers by personnel transfer, team building, and so forth.

Another learning point is the importance of close interaction and market orientation of the various functions in a business unit. Here, Asea was well ahead of Brown Boveri. The fact that Asea acquired a license from Matsushita while Brown Boveri decided to do its own research to come up with an improved product was another reason why the *time to market was much shorter* at Asea. On the other hand, Brown Boveri built up a sizable knowledge base over the years, making it possible for ABB to continue making rapid advances in new products and gaining more market share. A net present value calculation for the two approaches shows similar results (see Chapter 7), but it does not account for the value of the knowledge base created in Baden and used in both business units of ABB.

SUMMARY

We have seen a continuous change towards using more and more innovation to counteract the effects of cost reduction in established businesses and maintain a competitive position in a world with an increasing drive toward globalization. Resources, which until recently were not a bottleneck in manufacturing, are starting to become scarce.

The purpose of an organization is to serve many stakeholders, but the primary function is to develop and manufacture products. Making money is an important boundary condition, but this should not be the main driver. Various forms of organization are discussed, and a general trend away from hierarchical line functions to new forms of organization with better cross-functional communication and interaction has occurred.

While, in hindsight, each new product has to evolve in a stepwise process, which has been called the *linear model of innovation*, a forward-oriented view shows that reality is much more complex. Rather than going systematically step by step from one process to the next, continuous interaction and feedback are required to keep the time of development short and guarantee market success. A main conclusion is to move from the linear to a more-dimensional model and make "use-inspired basic research" the preferred mode of innovation. This conclusion should not be misinterpreted that pure basic research should not be supported. Basic research remains a very important part in innovation, but it is the task of management to look for continuous interaction between basic research and an understanding of newly emerging needs coming from the market.

The innovation process can be seen in three phases. Depending on the author, the first phase is the fluid phase or the fuzzy front end, followed by the transitional phase or new product development. The third phase is the specific or commercialization phase. One model to describe the first phase is the new concept development model. Knowing that this phase is still rather chaotic and dominated by a large number of unknown parameters, the model tries to provide a fairly complete overview of how innovation can be driven, which are the influencing factors, and which are the tools available to navigate among the five controllable activity elements: opportunity identification, opportunity analysis, idea generation and enrichment, idea selection, and concept definition.

The concept of the S curve illustrates the introduction, growth, and maturation of innovations as well as the technological cycles that most industries experience. One has to continuously look out for emerging technologies that may have to be introduced when they reach a certain level of maturity. In addition, one has to be aware of the possible emergence of a disruptive technology, which progresses independent of an established one. Once it satisfies the level of performance demanded by the established technology, the disruptive technology can invade, replacing the established technology.

The case study on two ways of creating the business of ZnO-surge arresters in two companies that later merged is a good example of use-inspired basic research at the research center of one company. It also shows the importance of working in a boundaryless organization and the importance of time. It also serves as a good example of the different thought processes going on in a research organization and a business unit that are not well connected.

REFERENCES

[1] J. Niosi, Fourth-Generation R&D From Linear Models to Flexible Innovation, Journal Business Research 45 (1999) 111–117.

[2] D. Field, Science—Process and Discovery, Addison-Wesley Publishers, Reading, MA, 1985.

[3] V. Bush, Science—the Endless Frontier, A Report to President Roosevelt on a Program for Scientific Research, United States Government Printing Office, Washington, D.C., 1945.

[4] T. Kealey, The Economic Laws of Scientific Research, Macmillan, New York, 1996.

[5] A.P. Speiser, Research at Brown Boveri, internal publication No. KL 5025 E, printed in Switzerland (7801-18000-0), Classification No. 0001.

[6] D.E. Stokes, Pasteur's Quadrant: Basic Science and Technological Innovation, The Brookings Institution, Washington, D.C., 1997.

[7] E. Mansfield, Academic research and industrial innovation: An update of empirical findings, Research Policy 26 (1998) 773–776.

[8] A.J. Salter, B.R. Martin, The economic benefits of publicly funded basic research: A critical review, Research Policy 30 (2001) 509–532.

[9] W.J. Abernathy, J.M. Utterback, Patterns of Industrial Innovation, Technol Rev 80 (7) (1978) 40–47.

[10] J.M. Utterback, Mastering the Dynamics of Innovation, Harvard Business School, Boston, 1994.

[11] C.M. Christensen, The Innovator's Dilemma, Harvard Business School Press, Boston, 1997.

[12] P.A. Koen, G.M. Ajamian, S. Boyce, A. Clamen, E. Fisher, S. Fountoulakis, et al., Fuzzy Front End: Effective Methods, Tools, and Techniques, in: A. Griffin, S. Somermeyer (Eds.), PDMA Toolbook for New Product Development, New York, John Wiley & Sons, 2000.

[13] P.A. Koen, et al., New Concept Development Model: Providing Clarity and a Common Language to the "Fuzzy Front End" of Innovation, Research Technology Management 44 (2) (2001) 46–55.

[14] P. Trott, Innovation Management and New Product Development, third ed., Pearson Education, Harlow, Essex, 2005.

[15] C. Schüler, Publication in 25 Jahre ABB Forschungszentrum Baden-Dättwil, Press Center ABB, Switzerland, 1992.

Application Phase—Design and Manufacturing

OBJECTIVES

The objective of this chapter is to explain some of the characteristics of the second phase of the innovation process, the application phase. This phase consists of engineering design, production engineering, and manufacturing. Each element is discussed in some detail, and the main emphasis is on manufacturing processes.

The many available manufacturing processes can be grouped in various ways. The objective is to guide you through the process of finding the right process quickly.

Two case studies address the difficult transition from R&D into the application phase. The objective is to show some of the factors to be considered to make this transition a success.

ENGINEERING DESIGN

Engineering design of components made from metals, ceramics, or polymers is a complex task that requires consideration of many interrelated factors, which often may not be compatible. The key player in this stage is the designer. This person is the important first contact of a materials scientist, and he or she remains in this role, unless the materials engineer can play the role or gradually become a designer. A designer usually comes with an engineering background, which enables him or her to quantitatively assess the value of a design, whoever created it. With each new design, the designer faces a multitude of factors, among which to carefully select the combination that best suits all requirements. In general, these design factors can be grouped into three categories [1]: functional requirements, total life cycle in design, and other major factors in design.

Functional Requirements

A component, to fulfill certain functions, has to meet a number of *performance specifications*.

79

Let us take as an example a component in a gas turbine, the turbine blade. The function of a gas turbine blade is to convert the kinetic energy of a flowing hot gas stream into rotation of the rotor. The blade, to perform correctly, has to have an optimized aerodynamic form, withstanding a combination of static and dynamic stresses up to a certain temperature. Both stress and temperature vary across the volume of the component. Furthermore, the blade is subjected to both corroding and oxidizing environments. Design software models make it possible to provide a complete picture of the distribution of all these life-limiting properties throughout the volume of the component. The materials engineer, when faced with these data, has to come up with the material that fulfills all the necessary criteria. During the process of continuous upgrading of the product, the material to be used has to be continuously improved or replaced by a new, better, although perhaps more expensive one. Risks always are involved in moving to a new material. If they are not covered in the performance specifications, then lots of money may be wasted by going in a direction that does not lead to a solution.

Because of the many accidents due to explosion of boilers in the 19th century and other incidents, a discussion started to establish standards for design.

A *standard* can be defined as a set of technical definitions and guidelines, instructions for designers, manufacturers, and users. Standards promote safety, reliability, productivity, and efficiency in almost every industry that relies on engineering components or equipment. Standards are considered voluntary because they serve as guidelines but do not, of themselves, have the force of law.

A code is a standard that has been adopted by one or more government bodies and has the force of law. Standards are a vehicle of communication for producers and users. They serve as a common language, defining quality and establishing safety criteria. Costs are lower if procedures are standardized; training is also simplified. Interchangeability is another reason. It is not uncommon for a consumer to buy one component in one part of a country to be connected to a second component purchased in another part of the country.

Total Life Cycle in Design

It is tempting to calculate the cost of a product strictly on the basis of materials and labor, plus allowances for overhead, selling, and profit. Especially for mass-produced products, such as consumer goods and cars, more costs have to be added to arrive at the total cost to the end-user, such as the cost of energy to operate the product, cost of maintaining the product, and cost of disposal of the product. In addition to the cost of the component for the producer and the user is the cost to the society at large. The relative importance of considering the cost to the user and society during design depends to a large degree on the type of product. In gas turbines, the societal cost is related to the economic impact of emissions, both short-term ones like NO_x and long-term contributions from CO_2. To a materials engineer, the concept of total life cycle must also include the total life cycle of a product's components and materials in the product. A product should be designed in such a way that the components can be reused, or the materials in the component can be recycled.

The question of producibility affects engineering design significantly. Moving from a two-part compressor in an industrial turbocharger to one consisting of only one part with three-dimensionally curved blades eliminated precision forging of one component and made it mandatory to produce the whole component by machining. The increase in efficiency of the one-part compressor had to be weighed against a temporary increase in cost, although an unforeseen trend toward more types of component but at a lower production numbers later confirmed the need to move to the more flexible method of machining.

Other factors that come into play are durability, or the intended lifetime of a component. Is it better to make a cheaper component that has to be replaced a few times during the operation of a machine or a component that, although more expensive initially, lasts the whole lifetime of the product? The question of life cycle cost is actually much more complex. Typically, a NPV (net present value) calculation is applied to alternative solutions, considering nonrecurring and recurring investment costs, scheduled and unscheduled maintenance costs, usage costs, and disposal costs.

Other Major Effects in Design

Several, more formal factors have to be observed during the design:

- **Knowledge base**. The most important factor refers to the state of the art. The designer should know as much as possible about already existing similar products. Special features of the existing product may be covered by patents. Patents become open literature after a transition time of several months to a year. Technical people who analyze patents carefully are often able to analyze the technological aspects of patents on devices and processes that may be applicable to other products. Often a process that has been patented is replicated in a research laboratory. After a short learning period, new ideas come up that may lead to a new patentable process.
- **Standards**. Many products have to conform to certain standards. An example would be the 220-V, 50-Hz standard for electrical equipment in Europe, and the 115-V, 60-Hz standard in North America. Other standards apply to the fuel-economy of cars, emissions of engines and turbines, and the safety of either the builder or the user of a product.
- **Human and aesthetic factors in design**. These factors involve the ease of handling of a device or its appeal to the customer.
- **Costs**. A product that meets the basic functional requirements has to be cost competitive. Whenever a choice exists between different materials, manufacturing processes, or designs, the least costly variant will be chosen, provided the basic functional requirements can still be met (for more details, see Chapter 7).

PRODUCTION ENGINEERING AND MANUFACTURING

Production Engineering

Production engineering involves an analysis of the complete product design to determine the way the whole process should be split into individual steps or operations. Decisions have to be made: What to produce in-house and what part of production to delegate to subcontractors? Once the process is laid out, specifications have to be made of the manufacturing methods to be used and what machines and tools are needed to perform the task. To come up with input data for cost calculation, the time required to perform a specific method is measured or calculated, and the skills of the persons to operate the equipment are defined.

Since a range of products may pass through the process, opportunities for process improvement have to be looked for continuously. This may lead to a rearrangement of production machines in the whole process or the purchase of more efficient machines. Through an interaction between design and manufacturing, a given product can be redesigned to make production possible at a lower cost. The easiest way to achieve this objective is to standardize the design, the tooling, and the manufacturing methods wherever possible.

Manufacturing

A vast number of existing, developing, and still to be developed methods transform a material into a certain shape with a desired functionality. Handbooks have been written to summarize extensively those processes that have been developed and proven their value. Ashby, in several of his books [1–4], ties together design, materials, and manufacturing process and translates this knowledge into practically usable software. Independent of the class of material, manufacturing processes can be put together in a systematic overview (Figure 4.1). Todd, Allen, and Alting [5] created a classification of manufacturing processes that is widely accepted in America. It distinguishes between shaping and nonshaping processes. The German industrial norm DIN 8580 [6] is similar. It is based on six main groups of processes, separated by the attributes of shape creation, shape changing, and nonshaping. Each main group has many subgroups. By partial regrouping (indicated by the group numbers on the lower part of the figure), we can distinguish among nine classes of materials processing: The so-called primary processes are those that convert the material into a shape. Most of the time, this is not yet the final shape, and is called *semifinished*, although there are exceptions, where the final shape is obtained after the primary process. The secondary processes are applied afterward to contribute to the shape change or apply specialized heat treatments to influence the microstructure and thus the properties of the material. Coatings or methods to change the properties of the surface are further methods. With the development of nanotechnology, these methods achieved a very important status.

Out of all these interconnected variables emerges the question, How do we select the right material and the right process for the design with the right function?

Even though all the available data, tied together in a nice software package, seem to make it a routine effort to come up with the right answer, new, unanswered questions always emerge as we move into the forefront of technology. We can go one step further: Following established guidelines inhibit the motivation to explore new territory. The following case study on isothermal forging of titanium impeller wheels serves as an example.

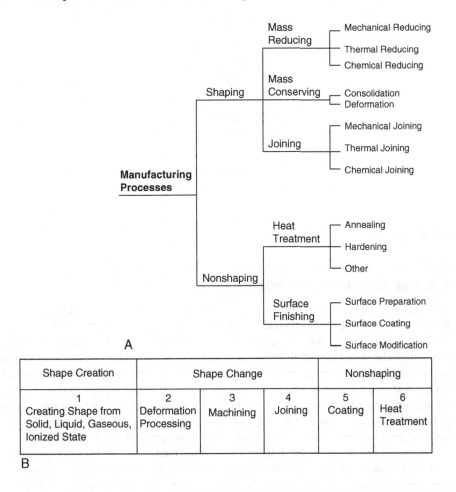

FIGURE 4.1 Classification of manufacturing processes: (a) Todd et al. [5], (b) DIN 8580 [6]. Reproduced by permission of DIN Deutsches Institut für Normung e.V. The definitive version for the implementation of this standard is the edition bearing the most recent date of issue, obtainable from Beuth Verlag GmbH, 10772 Berlin, Germany.

(Continued)

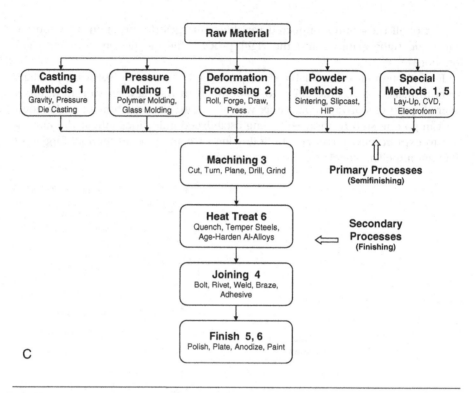

C

FIGURE 4.1—cont'd (c) primary and secondary processes (numbers refer to Fig. 4.1(b)).

Case Study. Isothermal Forging of Ti$_6$Al$_4$V Impeller Wheels: A Detour Toward Flexible Manufacturing

In 1972, one year after joining the Brown Boveri Corporate Research Center in Baden, Switzerland, I was contacted by two key engineers of the turbocharger division, who were then responsible for turbocharger development. They explained the function of the components of a turbocharger: the compressor and the turbine. As the general trend was to develop higher-power machines at lower cost and more compact sizes, they were considering replacing the existing aluminum alloy Al2618 with Ti$_6$Al$_4$V titanium alloy, which could resist higher temperatures. It was known that titanium alloys were much more expensive and more expensive to work and machine than aluminum alloys. The current state of design was to segment the impeller into two parts, a larger diameter part consisting of blades that were axially aligned and a smaller wheel section with three-dimensional shaped blades. The Al wheels were produced by upset forging. Depending on the size of the wheel and the quantities needed, they would be both forged and machined, or for larger production volumes, precision forged.

I was approached because of my know-how in powder metallurgy. They came to me under the assumption that this was the best method to manufacture future titanium wheels. The market input was that up to 8000 wheels per year would be needed in titanium, ranging in diameters from 180 mm to larger than 800 mm. That was the initial input. The first major mistake was that no one questioned the predicted sales or tried to make any effort to verify the data.

First Challenge: Available Manufacturing Methods

The challenge came at a time when new discoveries in high-performance powder metallurgy were made almost monthly, like powder atomization, rotating electrode process, and several methods of hot powder compaction and deformation [7]. Also, for nickel-base and titanium alloys, Pratt & Whitney in Florida, in 1970, had developed a process it called *gatorizing* [8], which allowed the slow deformation of high-temperature alloys in the superplastic state. Some of the information was openly available in published literature, but the patent literature, which was followed systematically, proved to be more important.

When learning about the high price of Ti$_6$Al$_4$V prealloyed powders compared with the price of titanium alloy billets, the need arose to think of alternative ways of manufacturing. One possibility would be to take a preforged billet and machine it to shape, as with the aluminium alloys. Machining titanium alloys, however, was a very expensive operation. This led to another idea: Why not use a variant of gatorizing to produce the impeller in net shape? This required a press, which we lacked, and the knowledge of how to deform at slow strain rates and constant, high temperature. At this stage, we had several options for manufacturing the wheels but no idea how to do it. Also at this stage, no cost calculations had been performed. The research director had no answers, but he helped to set up a three-day course where we learned about fixed and variable costs and the influence of volume on cost.

Quickly the following methods were singled out for comparison:

- Hot isostatic pressing (HIP) of prealloyed titanium alloy powder.
- Upset forging with either warm or hot dies, followed by precision machining.
- Precision forging at a low strain rate (superplastic deformation).

In superplastic deformation, the flow stress of fine-grained Ti6Al4V decreases several orders of magnitude at low strain rates and higher temperatures [9].

With the needed input about the yearly expected volume of Ti impellers (5000–8000), we quickly estimated that precision-forged Ti impellers using isothermal forging would have the lowest cost. HIP of powders and forging plus machining had to be eliminated.

We went on to make rough calculations, making assumptions about the cost of the machines, the tools, the time of processing, and so on and came to the conclusion that, for lot sizes of more than 50 wheels, isothermal forging was the lowest-cost method.

Second Challenge: Process Development

Information could be obtained from the published literature, but more important, from the patent literature. First, we had to learn about superplastic deformation.

FIGURE 4.2 300-ton hydraulic press used in the research phase of isothermal forging of titanium impellers.

One of the main inputs came from a technical engineer, who had become frustrated, felt misused by the Ph.D.s, and was ready to leave. When asked what he would do if he were given the freedom, he came up with an idea for a self-built press, consisting of a large diameter tube and a bottom and top plate to seal it off. Tension bars would compress the whole arrangement (Figure 4.2). Inside he would place the hydraulics for a press, and we would have a 300-ton press, which could be evacuated or filled with argon. The idea was approved, and soon the parts of the press arrived, shocking the boss: Will it work? The whole press would cost about 30,000 CHF (Swiss francs), a very large amount of money for the department at that time. We took the risk, and the press worked well immediately, because the design was very simple.

The first experiments involved setting up cylindrical specimens at constant temperature and various strain rates. This established the window for superplastic deformation. The next task was to learn how to master the forming of impeller-shaped geometries. To understand the variations of strain, we used model materials (lead, and later plastic, layered in various colors) (Figure 4.3). Progress, though, was quite slow.

We then hired a professional in metal working as project leader, and things started to move. With the help of two excellent technicians, we soon were able to produce 10-cm-diameter titanium impellers, using TZM molybdenum dies. After many trials and errors, we solved the most important problem: to find the

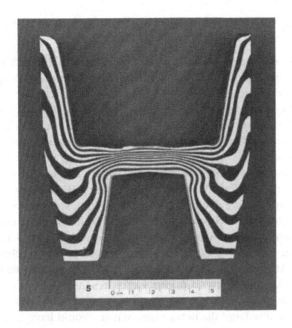

FIGURE 4.3 Modeling superplastic deformation with two colored layers of plasticine.

lubricant that would permit subsequent ejection of the forged workpiece. New knowledge was introduced by hiring experts on finite-element modeling of plastic deformation, allowing us to predict the strain rate at each volume element of the most complex shape. This would permit us to predict the total force required for the isothermal forging process. The reason for this complex simulation process was the high cost of the TZM dies and the need to prevent their breakage [10]. We now had all the computational tools available to predict the force required to forge any diameter of impeller and be certain that the die would withstand the stress.

The next steps were upscaling and technology transfer to the turbocharger division.

Third Challenge: Upscaling

1. **Ti billets**. We knew the range of impeller diameters. This gave us an idea about the dimensions of the Ti billet diameter. This was no problem; several suppliers existed.

2. **TZM die material**. This was a major problem. There was only one major supplier, in the United States, and there was some concern about an assured supply of the die material. A second supplier had to be found or built up. This was Metallwerk Plansee in Reutte, Austria, which we visited to inform the company about our intentions. The research director responded positively and helped set up the expensive, high-risk production. This decision was

facilitated by the fact that they were already dealing with the American market, so we would not be the only customer.

3. **Isothermal forging press**. We had to decide which type and size of press would be best suited for manufacturing. It was tempting to upscale the laboratory press, because it would be cheap—and it was our invention. I had visited the three largest U.S. forging companies, Ladish, Wyman-Gordon, and one company in Florida, which had the largest hammer press in the world. I was shown the actual presses and was quite scared about the size, which I translated into risk if we wanted to set up manufacturing in-house. But, we also contacted a German press manufacturer, who wanted to get into the business of isothermal forging presses. He impressed us by his professionalism, and we decided to work with him, starting with a smaller version (600 tons), which we set up in the lab. It was large enough to make the smallest diameter wheel (180 mm diameter; Figure 4.4). Reaching an agreement on the size of the production press to be purchased was more difficult. At a high-level meeting with the business area manager and the production manager, two options were presented:

☐ 3000-ton press: This would permit manufacturing most of the wheels, except a few large diameter ones, which would have to be machined.

☐ 8000-ton press: This would be large enough to cover all wheel sizes.

The business area wanted to get the 8000-ton press, but it finally agreed to go for the lower risk 3000-ton press.

FIGURE 4.4 180-mm-diameter titanium impeller, as forged.

4. **Outsourcing**. This question was never seriously debated. The business area, being the clear market leader, also wanted to maintain its position by in-house manufacturing.

Fourth Challenge: Technology Transfer

It was of utmost importance to successfully transfer the technology. Once transferred, the research center would have nothing to do with the day-to-day operation of the press. We faced two options. Either we would ask the most qualified researcher to move with the project into the business unit or we would get a specialist from the unit. We chose the second option. We asked the business area to send one of their best people to the research center for one year. First, that person would be taught everything we knew, reporting to the group leader in the research center. Toward the end, the business person would take command, reporting again to the business area.

The person was found quickly, a young and very eager man from the production department. He was assisted by a technician, who later became central in setting up the 3000-ton press and operating it.

After the press had been set up successfully, two things happened. The number of titanium impellers given to us by the business area was grossly off on the high end. Since we had a working press sitting around, we were asked if we could also isothermally forge Al-alloy impellers. We successfully implemented this technology in a short time, although it did not make economic sense.

The second development would prove to be the really devastating one: The designers came up with an improved version of Al-based impellers with three-dimensional shaped blades, a geometry that could be obtained only by machining. This version had been mentioned among the designers before the decision to purchase the press. Our mistake was that we did not participate in the internal discussion among designers and were satisfied when the decision was taken in favor of Ti-based impellers.

With these two new developments, there was no further reason to operate the press. Fortunately, we quickly found a buyer, Daniel Doncaster Ltd., and so ended this adventure in 1984. No major personal consequences followed the decision to stop the project, although the major players followed different career plans soon after.

More than 30 Years Later

1. Other things happened: The number of types of turbochargers increased, while the time to make them decreased by an order of magnitude. The requirement to machine complex-shaped Al-impellers remained. To further keep the cost down and get closer to the market, a joint venture with a Chinese partner-company was initiated. This confirmed, in hindsight, the correctness of the decision to withdraw from isothermal forging.

2. Metallwerk Plansee continued to produce TZM for dies and is now offering pieces weighing up to 5 tons.

3. The project leader became production manager in the business area, still remembering isothermal forging as his most exciting project.

4. Isothermal forging is still a process used by established companies. Ladish in the United States operates the world's largest 10,000-ton press; other companies have set some up, too.

Lessons Learned

1. Everything was done correctly to develop the isothermal forging process in-house, and we had the best possible team of experts to do the job.

2. Listen to new ideas coming from all members of your staff, including those from whom you might expect them the least.

3. Technology transfer was done perfectly, also the setting up of the 3000-ton press.

4. More effort should have been spent to check the market data. Brown Boveri, at that time, was clearly more technology than market oriented.

5. The two design options—titanium impellers, which could be made by forging, and the alternative, aluminium impellers with three-dimensional curved blades requiring machining—were superficially discussed early on. A critical review including cost analysis should have been done, involving people outside the group interested in developing the isothermal forging process.

6. Challenges 3 and 4 are not comfortable for an enthusiastic researcher, because they undermine and endanger a potentially exciting R&D project with important research content.

7. The last time to ask questions 3 and 4 was before ordering the manufacturing press.

8. Future question: Which is going to be the lowest-cost and fastest flexible manufacturing method? No one predicted that the market would demand more types of turbochargers and each type at lower numbers, moving production to the most flexible manufacturing. The business area, by quickly getting in and out of new technologies, demonstrated enough flexibility to remain the world leader in this business.

9. A former coworker, who had moved back to England and was now working for a forging company, approached us and bought the 3000-ton press, saving his company many years of costly research.

While this case study describes the evolution of a new technology from a materials science base, there are other possible starting points. The following case study describes the evolution of an engineering business from mathematical modeling.

Case Study. LNT: From Process Modeling in Russia to HIP Manufacturing Worldwide—Keep Focusing on the Content

Introduction

The establishment of a small highly specialized company to exploit innovative techniques in materials processing is the theme of this case study. Modern methods for processing materials, such as casting and forging, rely increasingly on complex computer modeling methods to define optimum processing parameters, thereby ensuring "right the first time" fabrication of complex shapes. This approach has largely replaced "trial and error" methods traditionally used until about 20 years or so ago. The development of computer modeling in materials processing was driven by the need to save cost by eliminating wasteful forging and casting trials and ensure consistency of properties in critical components and became possible as a result of the availability of personal computers (PCs) with increased computing power.

Background

Victor Samarov studied mechanical engineering at the best technical university in the former USSR (MPTI, Moscow Physical Technical Institute, a replica of MIT aimed at developing engineers with a university level of physics and mathematics). As is normal at his stage in life, he had no clear idea of where to go. His master's thesis dealt with a numerical method of describing crack propagation. He considered continuing his studies to get a Ph.D., but his professor advised him to go to VILS (where he had a friend, Victor's future boss), telling him that the company had just turned from a production plant to a research organization (with a plant), and he could find a multitude of tasks to be solved by process modeling. VILS was a very special place—the All Soviet Union (V) Institute (I) for Light (L) and Special Alloys (S), which was the central organization responsible for developing technologies for the Soviet aviation industry. It employed 3000 engineers and technicians in the institute and 5000 more in the plant located at the same place and had supervision over the other eight giant forging, casting, rolling, extrusion, and so forth plants.

The important point is that this institute was well equipped with a wide range of capital equipment, and this provided the opportunity to gain practical experience in the fabrication of complex wrought parts. Also, the strong team of talented, creative materials engineers gave Samarov a challenging technical environment where new ideas and concepts were encouraged. With his background in mathematics and physics, Samarov began to develop quantitative models, based on the physics of deformation and fracture in engineering alloys, to predict deformation during forging and other fabrication processes. The availability of extensive fabricating facilities provided the opportunity to validate his models, which is a key requirement for a reliable methodology.

A new development was the use of powder processing in the manufacture of high-performance components such as discs and blisks (turbine discs with integral blades fabricated as a single part) for military engines. The desirability of avoiding costly, time-consuming machining, which was often required for complex parts, led to the introduction of powder processing, and with process modeling, net shape manufacture became possible, which was a major breakthrough in fabrication from powder. Hot isostatic pressing was used to consolidate the alloy powder, and a quantitative knowledge of the complex processes involved in this technique, coupled with a powerful computing capability, enabled the reliable prediction of the design of capsules and the determination of processing parameters to give reliable net shape fabrication.

He had a good start, coauthoring a couple of articles on his R&D work and filing some patents. There was an abundance of talented and creative materials engineers, and Victor found the environment new and very exciting. Probably for the first time, he could see how ideas were born and very quickly transformed into processes and products. There was an abundance of money, everything was actually free, and one could do everything and anything one wanted at the large plant, with 5000 people and all kinds of equipment. This continued for one year, when the lab was transformed to serve the new direction of PM (powder metallurgy) superalloys and be responsible for developing the consolidation methods. Most of the older engineers were unhappy—they had been very comfortable within their old, often unique, knowledge—but Victor liked the new challenges. All was new, they had a lot of interaction with the other labs, and the leader of the new "direction" was enthusiastic and knowledgeable. All who worked there had a right to express themselves, and very quickly he found that he could contribute a lot with his knowledge and his vision of material behavior developed so far. Here evolved the first mathematical models of HIP describing the nonuniform shrinkage of capsules with powder and enabling prediction of their initial shapes.

My book, *Powder Metallurgy of Superalloys* [7], published in 1984, was translated into Russian. After meeting at a PM Conference in Dresden, we agreed to meet in Moscow. Victor had a team of about 10 engineers around him, working with him. He started to talk about starting a company.

Laboratory of New Technologies (Samarov, personal communication)

There was no vision, even no idea, to start a business before 1990, when the first signs of the Russian "free economy" showed up. In that environment, one could always find free helpers to do anything one thought was worthwhile. Thousands of small businesses appeared everywhere, perhaps hundreds in VILS alone, but they mainly resold the products generated or produced by large companies, taking away most of their profits. It goes without saying that it could not be done without unofficial approval by management, which openly or discreetly participated in all of this. The revenues were at the disposal of the small businesses, and in various forms, they gave back a part of it, but not to the large companies.

The original objective of the Laboratory of New Technologies (LNT) in 1992 was just to be a consulting company; nobody was planning to leave VILS, but it was practically impossible to provide such services through VILS. LNT was different, or considered itself to be different. It produced engineering products, and consulting was not the business of VILS. LNT had no bosses; they provided the collective consulting to Tecphy, ABB and many other companies all over the world, including Rocketdyne in the United States and research centers in South Africa and Korea. For example, LNT generated databases of rheological properties for PM alloys developed in the West and computer modeling of these materials and the shapes made out of them. LNT followed a kind of ethics: All jobs requiring the use of the equipment, materials, and facilities of VILS were not considered the product of LNT but solely of VILS. LNT also facilitated the first U.S. contracts for the development of impellers for rocket engines for Rocketdyne. This approach was also important to maintain confidentiality about specific Russian technologies, such as heat treatment and the chemical composition of alloys, although not too many people were interested anyway. This way, LNT did not share revenues with the management of VILS.

It all stopped in 1994, when the policy of the VILS management became extreme, and he openly told them about LNT, making him persona non grata for four years, after which he was even invited to become a member of the VILS Scientific Council. Now, LNT had to earn all its money itself. Consulting was not enough to provide the needed income, which meant that at least samples or pilot hardware had to be developed, and surprisingly, LNT was confident that it would be able to do it.

There were two reasons to leave VILS. Perestroika and the subsequent change to democracy gradually led (in 1994) to actual privatization, which often meant a robbing of the companies by its leaders. Employees got nothing; the worst took over power and leadership, which was disgusting to see and, of course, worse to participate in. The second reason came from several visits of Victor to the West in 1989-1992, where he could meet people, and to Moscow, which helped him to understand his own potential and that of the people working for him. His new partners and customers were interested in his knowledge and ability to conduct R&D to generate new knowledge.

Although he could not use VILS, cooperation with the French was in full swing, and they could do anything, using Tecphy as their first technology base. For several years, they even discussed a joint venture and had long-term royalty agreements, when the market in the United States began to unfold. LNT also tried to build a joint venture there, but the interests still were too different. The U.S. companies wanted to dominate because they were large and had production equipment. LNT thought that its new product ideas and tools would at least provide a balance.

At the beginning, just three or four people were employed by LNT, the others helped on a temporary basis as consultants. Later, when the volume of operations increased, the company had to build an internal structure and reached 15, sometimes 20, employees, but except for one accountant and one technician, all were first-call engineers. No other company in this field had such a concentration of brain power. In the beginning, most of the operations were managed through LNT Moscow, and most of the employees worked from there. When the U.S. market opened up and grew, LNT PM Inc. was created. With revenues now coming mainly from the United States, later also Synertech, a joint venture between LNT and the owner of Kittyhawk, a HIP company, was created.

Typical first jobs were in analysis, modeling, design, material tests (easily to be done at universities and other labs for cash), followed by manufacturing tooling and hardware, for which a small network of vendors was established. These vendors were mainly the best aerospace companies, desperately looking for jobs at the time. They submitted symbolic quotes, but all work participants preferred cash payments.

A bit later, when international cooperation started to unfold, to avoid licensing issues in the use of appropriate materials and have guaranteed quality and delivery, LNT started to use foreign vendors, at first the French, the first customers, later the United States. Eventually the jobs became more serious, making prototypes, which involved modeling, design, material testing, and the like.

The annual volume was $200,000–300,000 at the beginning, and more than half was paid as salaries for four or five people; the rest was needed for reinvestment and for overhead, to keep the company and travel.

LNT was "public" from the very beginning, although this was never advertised at VILS, and there were practically no contradictions. It is important to mention, that from the very beginning, LNT involved several people absolutely not associated with VILS, with their own expertise and ideas. LNT was officially registered, had a controlled account, reported taxes every quarter, and was also registered for international activity, as many contracts were with the West (and even with the East).

LNT already had enough revenue to buy computers and office equipment, and established a network of vendors who loved working with it. One of the six LNT coowners had a large apartment in the center of Moscow, but lived in the countryside, so the flat was made available to be used as an office.

Somewhat surprisingly, for many years, LNT had an official license from the Russian Aerospace Agency to officially participate in some government-funded projects (like SBIR in the United States). In France, LNT carried out work for France and the United States; in Russia, mainly for Russia. The core business was a contract-based work, and Victor's major task was to generate enough contracts and later more and more. This could be done only through permanent communication with the existing and potential customers, which revealed what they could need and where LNT's expertise could be applied. In addition to consulting, some pilot work was done: The Russian facilities were desperately lacking jobs, so LNT was always welcome; in France, work was based on lots of joint interests; in the United States, it worked in a common customer-vendor way. LNT did not invest in capital equipment, but used existing equipment elsewhere.

Sometimes, LNT paid (always in the United States); sometimes (especially in France), it was simply on account, as LNT did a lot of work there and a balance was derived. "LNT was profitable from the first minute."

Outside expenses, as just described, were not very high. The salaries were growing with the growing business from $5000/person on average to >$30,000/employee. LNT never had to use outside financing. For the first eight years (until it established Synertech), it did not want to have its own equipment; a joint venture with Tecphy in France (later Aubert & Duval) was discussed for several years.

Establishing Synertech

It became apparent that the burgeoning market in the United States could not be adequately serviced from Moscow, and it would be necessary to find a partner for a joint venture based in the United States. Eventually such a partnership was formed with a company in California, Kittyhawk, which had several large HIP facilities and worked closely with the aviation industry. The new joint venture, called Synertech, began trading in 1998.

The work load gradually increased, and eventually the new company became involved in the challenging task of prototype manufacture, which involved modeling, design, materials testing and the like; in other words, the type of work for which Synertech was ideally suited.

Since the early days, Synertech has gone from strength to strength and revenues have grown steadily. For the most part, the work being carried out is of the "one off" variety, involving a large amount of intellectual input to devise a powder fabrication technique that yields a net shape at a competitive price. In some cases, it was established that the design requirements could be met only by net shape manufacturing using HIP consolidation of powder. In this activity, the model developed to predict powder shrinkage is the key to the success of the overall operation. As better descriptions of the HIP densification process were developed and greater computer power became available, the model was refined and improved to ensure continuing high-quality output from the fabrication process.

The Russian LNT exists independently (for some work in Russia, where Synertech is not involved at all) and actually partners with another PM/HIP company. No legal links exist, but the partners carry out all the manufacturing work under favorable prices. Also LNT PM Inc., in the United States, continues consulting work, whereas Synertech does piloting and production work.

The Shrinkage Model

At first there was a simple model of shrinkage under isostatic pressure. Quickly, the company found that a capsule that provided an initial shape introduced substantial "plastic stiffness" and the resulting shrinkage was far from uniform. To make good predictions for more or less complex shapes, considering the high cost of powder and machining, prediction of the deformation field or capsule design was necessary. This generated the need to mathematically describe the deformation of a ductile (at HIP temperature) metallic capsule filled with compressible powder. To carry out such a calculation, it turned out that even a plastic model would be sufficient, as 90% of the deformation was plastic. This was LNT's main strength, as most of the scientists tried to introduce all possible mechanisms, which made their models completely impossible to use for real applications, since the database needed for each material would become astronomic, and the precision would still be bad. During these developments, Victor understood the importance of the HIP trajectory, as the relative strength of the steel used in the capsule and the powder materials differed during the HIP cycle. Here came the need for a better description of the HIP densification itself, involving powder particle rearrangement, plastic deformation, creep, and grain boundary diffusion. Later, everything was integrated into a single model, and LNT is now working on the eighth version to satisfy its internal needs.

Examples of Complex Shapes Produced by HIP

Component size is limited only by HIP furnace capacity. Many potential applications include dual-property, integrally damped turbine blisks.

Figure 4.5 shows an example of a Ti-shrouded impeller of an upper-stage rocket engine. If a five-axis impeller is machined from a forging in quantities of one to five, it is extremely expensive, about $80,000–100,000/part, the cost of the forging alone being only $2000–3000. On the other hand, if production by HIP of powder is chosen as a manufacturing technique, then a cost of $40,000–

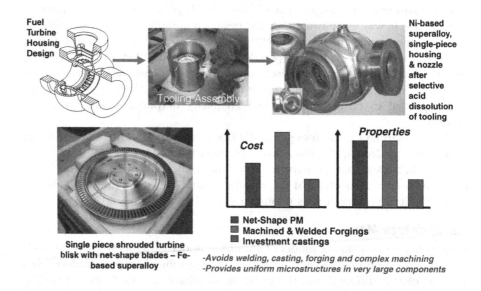

FIGURE 4.5 Selectively net-shaped shrouded Ti impeller produced by PM HIP.

50,000/part can be achieved, half the price of machining from a forged blank. The cost for about 10 kg of powder is about $1000–2000, the cost of the very expensive expandable tooling is $10,000.

In principle, most materials could be used in the HIP-consolidation process. Economically, most sensible are high-temperatures, high-cost, and difficult-to-machine materials. Typical examples are Ni, Ti, Be alloys, WC-Co, Ni-Cr-Si-B matrix, wear-resistant steels and alloys, duplex steels, and composites.

Competitive Cost Analysis and Market

Nobody really wishes to do a very detailed analysis comparing competing technologies, as people tend to protect their business and experts can never get a real estimate; in addition, the situation can change in a minute with a design change. Again, by manipulating the profit margins or doing some work in China, the estimate can be put upside down. So, it looks like the real numbers come only from looking at the competition.

There is also a correlation between optimal design and manufacturing technique. Very often designs are simplified so that they can be machined. The customer who finally understands often returns to the optimal nonmachinable aerodynamics that can be produced, together with the required mechanical properties, by HIP only. If a part cannot be machined, needs splitting and welding, and so forth (imagine a shrouded turbocharger), HIP can be 20–50% cheaper.

The total market for complex-shape PM parts (about $200 million) includes a part of those that are cast today and need improvement of properties and those that need design improvement and become nonmachinable. This comes in addition to those parts in the aerospace or off-shore industry already established via PM HIP route.

The main customers are in

- Oil and gas.
- Aerospace (mainly rocket propulsion).
- Electronics (sputtering targets).
- Power generation (gas turbines).
- Tooling for working hot and cold metals.

Reasons for Success

The main reason for the success of LNT, which still operates independently in Moscow, and the Synertech joint venture is undoubtedly attributable to the personality and technical ability of the founder of the company. Unusual for a top-rate scientist and engineer, he is also an excellent salesman with an unflinching belief in his own ability and a charm that puts colleagues and strangers alike at ease, even in difficult situations. For someone who grew up in the Soviet Union, he is remarkably at ease with people from the United States and Europe and is sensitive to the demands and the characteristics of the capitalist world. His drive and technical ability provided the basis for success.

However, to succeed in the development of net-shape manufacturing methods, which is the bedrock of the company, it has been necessary to fully exploit the potential offered by modern high-power personal computers, which now enable calculations to be carried out rapidly and cheaply on the desktop rather than with the large supercomputers that would have been necessary a decade or so ago. The ability to successfully exploit this capability has been a great achievement.

Future Outlook

In addition to continuously improving the software needed for modeling, more emphasis is to be expected on applying it to new classes of materials opening up new applications, for example, embedding long SiC fibers in a Ti matrix. A big remaining challenge is to open the market further for even larger dimensions, which would require HIP units of hitherto unseen dimensions.

Learning Points

1. The main driver behind LNT and the other ventures was a continuous interest to understand the adjacent and competing technologies and processes and perspectives of the market evolution and to continue to learn about the limitations of the process.

2. Regular business rules were hardly ever used to make decisions.

3. In spite of the many changes in the environment in which LNT was created, it was of great importance to remain focused on the original targets.

SUMMARY

In engineering design many boundary conditions have to be observed and met to make the development process successful.

To start with, a clear set of *performance requirements*, deduced from the functional requirements, have to be established. Next, all potential parameters that have an impact on the *life cycle* of a new or improved product have to be identified and analyzed. To avoid repetition in the development process, an extensive *knowledge base of the product's state of the art* has to be established. Since the product may have to conform to established *standards*, these standards and the compatibility of the product with them have to be known. Some products have to conform to *human or aesthetic factors*. Last but not least, a careful cost analysis process has to be set up to allow for a *cost comparison* between different variants of design and production.

Production engineering is the major next step, in which the transition from design to manufacturing is being planned. Decisions have to be made: Which part of the production will be done in-house and which part can be outsourced to suppliers? The complete series of production steps has to be known, and lots of thought spent on finding the most efficient way of manufacturing.

There is a large range of manufacturing techniques to choose from, and the choice of the right process will be driven by careful cost analysis and cost comparison between competing manufacturing processes. Ways to classify all available manufacturing processes in the United States and Germany are reviewed. Six main groups of processes are separated by the attributes of shape creation, shape changing, and nonshaping. By looking at the sequence of manufacturing processes applied to a starting material, which has to be transformed into a finished product, we can distinguish between primary processes, which lead to a semifinished product, followed by secondary processes, which lead to the final shape. Coatings or methods to change the properties of the surface are further methods. With the development of nanotechnology, these methods have achieved a very important status.

The case study on isothermal forging of Ti-alloy impeller wheels for turbochargers is a demonstration of perfect project execution on the technical side, from basic research to the design and operation of a new 3000-ton press. Nevertheless, two unexpected factors forced the project into a different direction. One, a newly designed impeller with a geometry that required a different manufacturing process. Two, the changing market needs away from a few types manufactured in rather high volumes to a large number of types but lower volumes. The company, however, took the right decision to change course, at the right time, remaining successful from the business side.

The case study on computer-aided design of preforms for hot isostatic pressing of powder shows the successful development of a new business because of the convergence between the design of a new mathematical model with the onset of high-speed computing, the availability of a "free" state-run R&D organization in Moscow, and a pioneering entrepreneur, who never lost track of his goals, to make his mathematical models perfect in predicting the shape of the starting geometry, which is needed to get a high-precision product after hot isostatic pressing.

REFERENCES

[1] M. Ashby, K. Johnson, Materials and Design: The Art and Science of Materials Selection in Product Design, Butterworth–Heinemann, Oxford, UK, 2002.

[2] M. Ashby, H. Shercliff, D. Cebon, Materials: Engineering, Science, Processing and Design, Butterworth–Heinemann, Oxford, 2007.

[3] M.F. Ashby, D.R.H. Jones, Engineering Materials 1: An Introduction to Properties, Applications and Design, Elsevier, Boston, 2005.

[4] M.F. Ashby, Materials Selection and Process in Mechanical Design, Butterworth–Heinemann, Oxford, 1999. DIN-Norm **DIN 8580**

[5] R.H. Todd, D.K. Allen, L. Alting, Fundamental Principles of Manufacturing, Industrial Press, New York, 1994.

[6] DIN 8580: 2003-09, Beuth-Verlag GmbH, Berlin, 2003.

[7] G.H. Gessinger, Powder Metallurgy of Superalloys, Butterworths, Oxford, UK, 1984.

[8] M.M. Allen, R.L. Athey, J.L. Moore, Metals Engineering Quarterly 10 (1970) 20.

[9] A. Arieli, A. Rosen, Superplastic Deformation of Ti-6Al-4V Alloy, Metall. Mat. Trans. A 8 (1977) 1591.

[10] C.R. Boer, N. Rebelo, H. Rydstad, G. Schröder, Process Modelling of Metal Forming and Thermomechanical Treatment, Springer Verlag, New York, 1986.

Managing Technology

5

OBJECTIVES

The objectives of the chapter are to show why managing technology is a continuous process to preserve tension between the resistance to change and change itself. Many parameters are subject to unpredictable changes, and they are discussed in detail.

Managing R&D also entails finding the right organization. Several concepts are discussed, and ABB Corporate Research is used as a case, showing the search for continuous improvement on many levels, mirroring similar developments worldwide.

MANAGING CHANGE

In Chapters 1 and 3, the image of a funnel was used to describe the flow of an idea from innovation from its source to launch to market. It is easier to draw such a picture in hindsight. The bigger challenge is looking forward and learning how to find the right course. Many parameters come together, and we always have to be ready to readjust the course because an unforeseen change occurs. Managing technology is an important element in managing business, and the most important thing to learn is how to anticipate and manage change.

Let us start with the bigger picture (Figure 5.1) to get a top-level view and go into detail afterward.

The first challenge in managing technology is to define and follow the right strategy of coming up with a product innovation.

Portfolio Management

R&D is vital to the survival of a company; it is an investment in its future. While few question this, the art of managing technology is to make the right choice from a portfolio of projects that should be supported. Experience has shown over

101

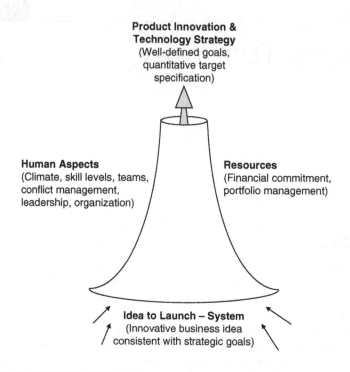

FIGURE 5.1 Multiple aspects to consider in management of technology.

and over again that there are no golden rules to help you make the right choice. Many managers in well-established, large enterprises tend to support primarily those projects that assure continuity and stability of the existing business and often are skeptical toward projects that promise radical innovation. Steele [1] rightly describes the need to preserve the tension between the resistance to change (e.g., support of the existing technology) and change itself, be it evolutionary or revolutionary, the most important challenge in the management of technology. The probability of success of new technologies often is not even 5%, while the probability of failure may be 95%. On the other hand, most publications focus on the few successes of technology development, creating a large pressure on management to support any new technology that has the potential of radically changing the business world. This trend to position new, high-risk technologies too high at the very beginning of a research effort has increased significantly over the years.

The challenge in managing technology is to be sure not to overemphasize support of the highest risk projects but also not be too conservative and support only those projects that yield incremental improvements with a high likelihood of success. This is a real dilemma for a manager, considering this person will be right 95% of the time when refusing to support a radical project with a potential high business impact but containing high risk. On the other hand, very often this could be one of the 5% of projects that constitutes the future success of a business. But very often, disruptive innovations are the domains of startup companies, which is also documented by some of the case studies in this book (NDC, Metoxit, Amroy, AMSC are typical cases; LCD is an exception, although the pioneering companies did not benefit from being first to introduce the product in the market).

Although the rewards for successful innovation can be spectacular, the real need for innovation is the survival of the company. We have to realize that each product has a finite lifetime. Sooner or later another innovation is needed to replace an existing product by a new one, although it is often difficult to predict when this will happen, because we cannot predict the future or when a competitor will come up with a new and better idea. We can thus see two tasks in technology management: supporting the existing product and coming up with new solutions to replace the existing product.

Innovation can occur on many levels, which Table 5.1 exemplifies.

Innovation is always coupled with risk and uncertainty. Depending on the location where innovation is taking place, these may differ widely. The highest risk can be allocated to product innovation, because here we find a combination of two risks: the risk to attain the required technical performance and the risk to find market acceptance. Process innovation, though important, is often invisible to the customer, unless it is coupled with cost reduction or improved quality. Incremental improvements bear the least amount of risk. Major innovations, such as LCD, are not only beyond the scope of small companies but even large ones, if they are positioned wrongly in the market. Major new systems or products can be introduced only by close government-industry cooperation, which by definition brings new risks. Examples would be photovoltaic products, wind power, high-temperature superconductors.

Table 5.1 Spectrum of Innovation: Examples of Different Types of Innovation

Innovation	Example
Incremental improvement in a product, process, or system	Higher-efficiency Al-impellers for turbochargers
New component, process, or technique in a larger system	High-risk/high-impact components in engines
New product for existing market	Nano-coated textiles
Radically new product	CNT-reinforced epoxide
New product for a new market	Stents made from shape-memory alloys
New system	Fiber optic communication
Entirely new capability	Nanotechnology in IT applications

The potential impact of new innovations is often distorted because of a select few big winners, which then are taken as synonymous for any innovation. In any large organization, it is quite common to perform business impact calculations on new projects. While this has the advantage that it forces the project leader and the manager on addressing several issues about the project, they have to be realistic that many projections in the future can be totally off in either direction.

Looking backward, we always pick the breakthrough innovations as a model that should be applied to projections into the future. As mentioned earlier, it is highly unlikely to predict a breakthrough in innovation, although we should always be ambitious about setting targets. More important and more likely to succeed is a continuous focus on incremental improvements. Many managers, though, might give it an undeservedly lower status.

Financial Commitment

Managing a portfolio of projects can also be seen as the challenge of how to optimize the return on investment into new technologies. An important part of a technology strategy is the financial commitment.

Does More R&D Expenditure Lead to Higher Profitability?

Collier, Mong, and Coulin [2] show a statistical correlation between R&D intensity (total R&D costs/revenues) and profitability, measured as the return on investment. Figure 5.2 is a simplified version of the results, showing growth curves of

Sales of new products/total revenues versus R&D intensity,
Market share of new products versus sales of new products/total revenues, and
ROI versus market share,

from which the growth curve ROI versus R&D intensity is deduced.

This statistical correlation is valid only if there is a causal relationship between R&D and new products. For example, it is important to allow for the time required to bring products to market. Many large companies do not go through the effort of following the long path of a research project from start to market success and, therefore, may correlate a large R&D effort with a successful business, even if there is no causal relationship.

These statistically obtained data should not be interpreted wrongly: Spending more on R&D might be a barrier to innovation, if R&D makes management forget that numerous additional parameters have to be considered to really succeed. In each phase of an innovation project, *all functions in the company have to be engaged*:

- Management must have a product market strategy, a plan showing where it wants to be in five years.
- Finance must provide necessary funding in the budget.

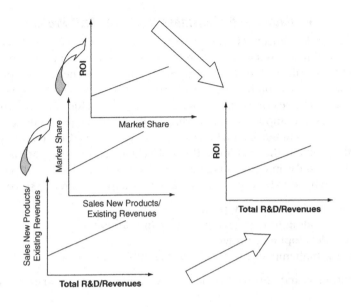

FIGURE 5.2 Simplified evolution of a graph showing the growth of ROI as a function of R&D intensity, based on data by Collier et al. [2].

- Marketing and Sales must analyze the market and the competitors, resulting in target specifications, and provide customer feedback.
- Production must be involved early-on in transferring the results of R&D.
- Quality Assurance must have all the required test equipment and test plans.

Zeller [3] gives a good example of an incomplete target specification for a project in ABB's power semiconductor business unit. A new project was under discussion, a gate turnoff thyristor with 9000-V blocking voltage and 2000-A current, which could be switched off. The technical department had come up with all the specifications, together with the main customers. Subsequent computer simulation made the target data look realistic. Estimated development costs were 400,000 CHF and an estimated annual sales volume of 2 million CHF was predicted. Everything looked fine and all signals were for *go*. The project, however, was stopped. The showstoppers were the estimated costs of 5 million CHF for required testing devices (12 kV, 3000 A, $dI/dt >$ 3000 A/μsec).

The example shows that a list of functional specifications sometimes is not enough. Typical gaps, which may be overlooked, are reliability, maintainability (examples are steam generators in nuclear power plants, and the combustor problems in a newly developed gas turbine), product life cycle, product design, and test plan.

Tools for the Calculation of the Business Impact of R&D Projects

A perfect metrics does not exist because of the diversity of the projects and the difficulty to accurately estimate costs and future returns. Still, there is some value in trying to come up with an estimate, which of course needs continuous checks and updates. Various formulas exist, beginning from very simple ones neglecting the time value of money, to more complex ones that consider both the time value of money and quantitative risk factors. It is important to note that the quality of the results depends on the quality and level of communication between the people involved, the complexity of the system or the problem, the duration of the project, and other factors. Although the formulas used are similar in other companies, we look at some that have been adapted from those used by ABB. First we have to distinguish among different types of projects:

- Projects leading to a new product.
- Projects leading to cost savings in existing products.
- Process development projects.
- High-risk/high-impact projects requiring a business plan.

Different formulas and approaches have to be used to accommodate the different types:

For *product development projects* (new products or a new generation of products including projects for lowering product cost),

$$BI = AR \times RF \times AM$$

where

BI = business impact.
AR = sum of five years additional revenues after the project is finished.
RF = risk factor in percent probability that the additional revenue will be realized.
AM = average margin of earnings after financial items for applicable product or type of products.
Here is an example: suppose you are developing a new type of transformer based on new materials. The yearly revenues from the existing product are $100 million. The new product will lead to M$20 additional revenues, raising the yearly income to $120 million. The probability that the R&D project will lead to a success is 70%. Then the business impact of this project, assuming a 12% margin of earnings, is

$$BI = 100 \text{ \$ million} \times 0.7 \times 0.12 = 8.40 \text{ \$ million}$$

For *process improvement projects* (non-product-specific development related to lowering cost for processes, e.g., engineering software development),

$$BI = CS \times RF \times 0.5$$

where

BI = business impact.
CS = sum of five years cost savings after the project is finished.

RF = risk factor in percent probability that the cost savings will be realize The factor 0.5 implies that only 50% of the cost savings is to be accounted for.

Mature businesses have to focus heavily on cost cutting to keep their products on the market. Thus many research projects concentrate on savings derived from process improvements. Very often, such projects are easier to assess, because a commercialized product already exists.

The advantage of both methods is that they are simple and fast to do. If done by experienced, business-oriented researchers, they provide a sense of the importance of a new project proposal. No business plan is required, which, as experience has shown, very quickly leads to grossly overstated values, especially if no regular follow-ups are mandated. The methods do not take into account the time value of money and thus give distorted results. Discussion within groups of people and across different programs help bring some reality into the figures. The main benefit of the method is that, very quickly, many different projects can be compared to each other and the more important ones selected.

The disadvantage is that often unrealistically optimistic predictions of future revenues are made, which make the project look more favorable

The *high-risk/high-impact calculation model* is the best model, but it also requires a great deal of effort to set up. For most research projects in a very early phase it contains too many assumptions to be worthwhile. It is based on setting up a business plan with projected revenues and costs over several years. Table 5.2 shows an example of such a model.

Total R&D costs are defined as costs accumulated up to the first commercial demo installation (i.e., excluding costs for engineering design, marketing, manufacturing setup, large prototypes, plant). IBT is defined as income before taxes. The formula for the (product development) business impact ratio (BIR) is

$$BIR = BI \times BP/TotR\&D$$

where

BI = business impact (sum of five years of IBT).

BP = business success probability, assuming technology proves successful.

TotR&D = total R&D costs with and without third-party funding.

This definition of business impact BI differs from the one used in the previous 2 formulas by focusing on total revenues rather than 'additional revenues after the project is finished'.

Another formula describes the return on investment (ROI), where, more realistically, all costs arising from research to commercialization and necessary new investments are included:

$$ROI = BI/(TotR\&D + Commercialization\ costs + Investment\ costs)$$

Table 5.2 Business Impact Calculation Model Used in High-Risk/High-Impact Proposals (in $ millions and %) [4]

	2000	2001	2002	2003	2004	2005	2006	Cumulative	%
Market size	220	220	220	220	230	235	240		
Market share	1%	5%	25%	43%	47%	48%	50%		
Revenue	3	12	54	95	108	113	119	503	
IBT	14%	15%	15%	16%	16%	17%	17%		16%
	0.4	1.8	8.1	15.1	17.3	19.2	20.2	82	
Corporate R&D funding	2.4	—	—	—	—	—	—	204	41%
BA/BU R&D funding	2.5	0.9	—	—	—	—	—	3.4	59%
Total R&D funding	4.9	0.9	—	—	—	—	—	5.8	100%
Cash flow	−4.5	0.9	8.1	15.1	17.3	19.2	20.2		
Cum. cash flow	−4.5	−3.6	4.05	19.5	3.68	56	76.36	76.3	
Commercial. costs	0.2	0.2	—	—	—	—	—	0.4	
Investment in plant	0.3	0.1	0.3	0.6	—	—	—	2.3	
Net cash flow	−5.0	−5.4	2.4	17.0	34.2	53.4	73.7	73.7	
Third party R&D funding	0.4	0.4	—	—	—	—	—	0.8	

Assuming

BP = 90%, the following results will be obtained from the numbers in
Table 5.2.
BI = $82 million.
BIR = $12.7 million (without cost of third-party funding)
BIR = $11.2 million (including cost of third-party funding)
ROI = 9.8%.

The methods described so far are oversimplified, because they do not allow
for the time value of money: *One dollar earned after one year is worth less than
one dollar now if discounted back to the present with an assumed interest rate.*

The *net present value* method uses discounting methods to calculate the present value of money spent or earned in the future. The method is described in
detail in Chapter 7, including a case study, where the NPV of a product development project using internal R&D is compared with the NPV of product development based on licensing.

Existing tools for making R&D investment decisions cannot properly capture
the *option value* in R&D. Since many new products are identified as failures during the R&D stage, the possibility of refraining from market introduction may add
a significant value to the NPV of the R&D project. The option value depends on
the expected number of jumps and the expected size of the jumps in a particular
business. For further details and critical comments, look at the literature [5,6].

Barriers to Innovation

The evolution of an innovation always has to be linked to risk and uncertainty. There
is the uncertainty whether a new idea will succeed and lead to a new product as
well as the uncertainty whether customers can be found and induced to use the
innovation. Most customers already use available, proven technology, and they face
a risk when they bet on a new idea that might turn out to be a failure, backfiring on
them. This uncertainty generates barriers that have to be overcome.

Steele [1] distinguishes between intrinsic and extrinsic barriers. *Intrinsic
barriers* relate to the fact that innovation requires a multitude of skills: the skills
of the inventor, the skills of the person who transforms an idea to a practical solution, the skills of the project leader, a person with marketing skills, and persons
with knowledge about financing. The case study CERCON® turned out a successful innovation because many skills were combined in a few individuals.

Extrinsic barriers are both managerial and organizational. Managers often
have to create certainty in their operations and might view an upcoming innovation as a disturbance of certainty. It is therefore the task of a CEO to emphasize
the importance of innovation to these managers—and many do. Still, managers
are faced with the problem of diverting some of their best people and important
resources to a new innovative project, despite needing the same resources for
less risky opportunities. Managers therefore require the skill to identify and

quantify the comprehensive benefits that an innovation might create. One can even go a step further and include second-order effects that could make an innovation's impact much greater. It is the task of a successful technology manager to understand where a business unit manager is coming from and find the right language to successfully communicate his or her plans. Managers face other stakeholders, each of whom comes with different expectations: employees want opportunity for growth in responsibility and income; shareholders and investment analysts focus on growth of a company's profits, and they like to see predictability of operating performance; customers want products with the latest technology and compatibility between generations of products.

In general, barriers are useful to have. Disruptional costs associated with innovation would be counterproductive if they did not lead to substantial improvement. Tough entry barriers help ensure that the innovations that succeed do indeed justify the disruption they create.

Any investment in R&D carries a risk. These risks can be minimized by bringing together all the functions of an organization and going through a complete checklist of things to do. Prior negative experience often helps as much as positive experiences to look out for the right barriers. Assuming that all these factors are known and thought through makes higher investments in R&D one of the most important contributing factors toward company profitability. Conversely, relying only on statistically obtained correlations almost certainly will lead to failure.

Side Benefits of Innovation

Although the declared intention of R&D is to come up with a new innovation, there can also be a defensive view. A company that is the leader in a certain product area has to prove that many alternative solutions do not endanger the existing product. The side effect of such attempts is that the conventional technology remains competitive. The work on high-temperature semiconductors by large electrical engineering companies serves as an example.

Further benefits of innovation are that it can be considered very helpful as a management learning experience. Starting something totally new is quite a different management challenge from keeping existing technology competitive by incremental improvements. The management team working on something totally new is exposed to the limits of managerial skills and decision-making judgment. We can also say that innovation prevents too much comfort, which can happen when success has been established. As I experienced early on in my career, it was not always easy to cross the hurdle of the arrogance of experienced managers, who were convinced they were following best practices and did not have to consider looking out for anything new, unless it was their own idea. A positive example of this argument is the case study on isothermal forging (discussed in Chapter 4). The project started out as an ambitious attempt to introduce a lower-cost manufacturing technology. After the project was stopped, a new approach towards highly flexible manufacturing was taken. The project leader of the forging project became production manager, and the business unit now has better experience than any of its competitors about the advantages and limits of isothermal forging for their business.

Human Aspects in Supporting Innovation

Successful innovation requires different types of support at the various levels of an organization.

Commitment

Innovation needs dedicated commitment from a number of players at all levels of an organization. Senior management and even the CEO can make a huge difference in motivating the innovation team, because top management can communicate on a higher level what the impact of an innovation may be for the enterprise. The researchers get the impression that their work is so important that even busy managers take the time to learn about it and think ahead about the likelihood of future successes.

People Involvement

An innovation needs a champion who can devote all his or her energy to one purpose, making the innovation a success. To extend this statement to the whole project team means that, if possible, this team should be completely detached from all other activities and commitments. People who get additional operational activities are too easily diverted, because it is easier to judge operating than innovative work.

Resources

Quite often the availability of resources is considered by many staff members as a major bottleneck in innovation. Funding of research projects from external sources (government, private associations) can often lead to a shift of focus: Acquiring as much external funds as possible becomes more important than focusing on the long-term business value for the corporation. Many of the externally funded projects have limited strategic value, therefore, and should be stopped. ABB, starting at a very high management level, communicated one slogan well, describing best the whole situation: "The problem is not the lack of money—the problem is the lack of good ideas."

Once this statement was understood by researchers, it served as a motivation to focus their thoughts on strategic contents rather on the availability of funds, and it became very easy to get funding from within the corporation from the business units affected by a new idea.

Accepting Failure Helps an Innovation

In operating management, where everything can be defined clearly, missing a target or failing to stick to a budget is evidence of incompetence. This attitude is opposite to what is needed to help an innovation. One has to accept the probability of failure as something inherent in the process of innovation. In the early stages of the Brown Boveri Corporate Research Center, many business unit managers were reluctant to accept results coming from the research center; because of a frequent lack of communication, they primarily saw the inherent risk. Rather than being proactive in supporting a new idea, they often were overcritical, which could significantly slow the process of innovation.

Implementation

Assuming all the above-listed commitments to innovation exist, how can we best implement it? Basically, there are three possible operating principles:

- Centralized approach: Innovation is a corporate-level function.
- Decentralized approach: Innovation is the responsibility of an operating function.
- Multilayer approach: Innovation is the responsibility of both corporate and operating functions.

Centralized Approach

Leaving the main responsibility of innovation to corporate-level functions is a very risky undertaking, and it becomes riskier the larger the corporation. At big pharmaceutical companies, the solution is often to buy the smaller innovator, especially in the biotech business. Innovation covers the whole spectrum from incremental improvements to innovation focusing on growth and diversification. Incremental improvement requires very detailed knowledge of the existing product and technology, which mostly resides in a business unit. Delegating it to corporate functions almost certainly leads to loss of focus and timely delays in project execution. The centralized approach aims at larger discontinuities. The advantage would be that it demonstrates corporate management's commitment to innovation, and it motivates management to show initiative to start new projects.

A possible disadvantage of this approach is that senior executives, obsessed with the challenge of growth but often not well informed about the critical issues of a project, begin to exaggerate the potential of the innovation so that it can have an impact on the entire business. Very large resources are put into the project, because everyone has been talked into believing that the project has much higher value than predicted by its originators.

For all its problems, many times a centralized approach helps overcome the resistance of uncertain managers and may be the only way to introduce a high-risk innovation, but it takes very competent top management.

Decentralized Approach

In the decentralized approach, the responsibility for innovation is assigned to operating managers of business units. On the one hand, it has the advantage of bringing many players familiar with the business into the picture. On the other hand, it requires operating managers open to managing the uncertainty that comes along with projects of a high innovation potential. It also creates a challenge to corporate management, which may have become accustomed to managing operating performance and now have to supervise from a distance higher-risk projects, which may require funds that exceed the money available to smaller business units.

In numerous cases, operating managers showed the right kind of leadership. One example was the issue of how to participate in the European COST-program, which provided modest financial support to high-risk research projects and the opportunity to team up with both competitors and suppliers of new materials. The operating manager stated very clearly and convincingly that it is the business unit that understands best the market and the potential benefit of a new high-temperature material to be used in gas turbines. He suggested the creation of work packages by the participating companies, keeping the funding agencies at some distance. The approach was immediately accepted by partnering companies and ABB was asked to assume leadership for the work package. One particular work package concerned the introduction of newly developed ODS (oxide dispersion strengthened) nickel-based superalloys for application in gas turbine vanes. If successful, the new high-temperature alloy would save costly cooling of the turbine vanes, permitting higher operating temperatures than were possible using cast alloys. Corporate research had solved virtually all the open questions: how to use isothermal forging of initially fine-grained, as-extruded bar material to get the final shape and developed a zone annealing process that gave the required elongated coarse-grain structure. Many companies participated enthusiastically in the project. Unfortunately, the project had been started when the size of industrial gas turbines was still quite small. With the development of combined-cycle power plants, ever larger gas turbines were required, which also meant that the size of vanes and blades had to be larger. At a certain stage, this size was too large to make extruded bars of a large enough diameter, from which vanes could be forged. Being very close in contact with operational management of the business unit, it was a very easy and quick decision to cancel the project, once this hurdle had been discovered. In hindsight, these alloys could have been introduced to small engines or considered for totally new applications. In a purely centralized approach, the researcher may have persisted in pursuing other objectives of the project and neglected the critical question that no equipment existed to extrude large diameter bars.

Another good learning experience was the setup of the Power Plant Laboratories at Combustion Engineering, then part of ABB. The research lab participated openly in working together with other corporate research centers, but it was the only lab managed directly by the business segment. Although, from a managerial point of view, this was a fairly small difference from a corporate lab, it brought operations much closer to the business units than in a corporate lab and much effort was spent making the business unit managers feel like co-owners of the research lab.

Multilayer Approach

In this approach, we seek the best of two worlds: sharing responsibilities and commitments on both the corporate and on the operational level.

An example is ABB's high-risk/high-impact projects. In the yearly budget presentation by the chief technology officer to the CEO, the CEO asked the question, "What would you do, if I gave you another $10 million?" This led to extensive brainstorming in the R&D community across the whole corporation, and the result was to create projects where the idea was created at the business unit but could be done jointly with corporate research and partially funded from corporate funds. A business plan was required and the project would be supervised by a team of managers coming from both the corporate and business unit levels. One success story for this approach was the creation of an entirely new product, the Azipod permanent magnet motors attached to the rudder of a ship, allowing very sharp turns. There were many other successes, and it was soon accepted that this approach was the best to promote a business-unit project or speed up the transfer process from corporate research into the business unit.

Guidelines for Success

In establishing a project to pursue an innovation, some guidelines should be followed:

1. Start with a small effort to allow for flexibility. This is the phase where you have to define a complete set of quantitative *target specifications*, show that in principle the concept is sound, and find out where the bottlenecks are. If you step in too big in the beginning, your efforts may be focused on only a part of the issues, and you may forget some important ones, which may show up later, after much money has been spent or wasted.

2. Make the project leader and the team aware that they are owners of the project. Be ready to manage conflicts of objectives.

3. Make sure that the *goals are well defined and understood*. The innovative business idea has to be consistent with strategic goals of the business. Provide good continuous guidance to achieve goals, and make the increase of financial support dependent on the progress achieved.

4. *Bring together all the functions* in all phases of the project team from the beginning, and make sure they combine all the required skill levels. This also includes outside suppliers and original equipment manfacturers as part of the value chain.

5. Create a *well-defined business model*, and obtain market inputs early and continually.

6. Get a *first application* as early as possible to test if your market assumptions were correct.

7. Move your project through predefined *go/no go gates*, applying change management.

8. Plan well ahead for the need for more financial resources.

By definition, this list will never be complete, as each project has its unique history and evolves in a different environment.

MANAGING R&D

In basic research, the role of the scientist is to look for a basic understanding and theories to explain facts. Engineers, on the other hand, usually look for technical solutions good enough to solve a customer's problem. Thus, basic research often is under central guidance and long-term oriented, while engineering development is under the guidance of the business unit and short-term project oriented.

While the focus in basic research is on developing knowledge, ideally of use in later product development, engineering focuses on developing a plan, consisting of drawings and details of processes required in the introduction of new products in the market.

As discussed earlier, in use-inspired basic research, some of the two elements have to come together right from the beginning.

Sites for Research and Development

Historically, the location of R&D of large firms used to be at the headquarters of the organization, an indication for the importance given to it by corporate management in a centralized organization. Over time, more and more dispersion occurred, for various reasons. While research remains at central locations, close to knowledge clusters, including top universities, development follows manufacturing and evolves from local market support or potential new markets. R&D has to combine the drivers for the two functions. As a result we have seen more and more decentralization of R&D to the main regions of a global company's business. A typical example is the explosion in the numbers of R&D sites in addition to lower-cost manufacturing sites in China.

Management of Interface Between Research and Development

As discussed earlier, research often starts with a multitude of ideas, many of which have to be discarded at a later stage, until the winning ideas survive and lead to the successful development of a new or improved product. The interface between research and development plays an important role and has to fulfill three functions (see a more detailed discussion of the new concept development model in Chapter 3):

1. In *project creation*, the most challenging phase, the right questions have to be asked.

2. To ensure that the right ideas are selected, a *filter mechanism* is used, in which project proposals are economically evaluated.

3. In *transfer*, the right ideas are properly implemented.

Organizational Structures

As discussed earlier, many options are available for organizing R&D units: centralized, decentralized, and multilayered or cross functional. The last brings together persons from different functions at an early stage, making sure that no boundaries exist to impede technology transfer from research to the business unit.

The best overview of how global companies organize their R&D activities is given by Boutellier, Gassmann, and Zedwitz [7]. Global organizations have to consider additional complexities in managing their research activities, as shown in Figure 5.3. One can distinguish four levels of structure in international R&D organizations. The most informal first level recognizes informal networking, which has gained substantial interest and importance through the IT revolution. The

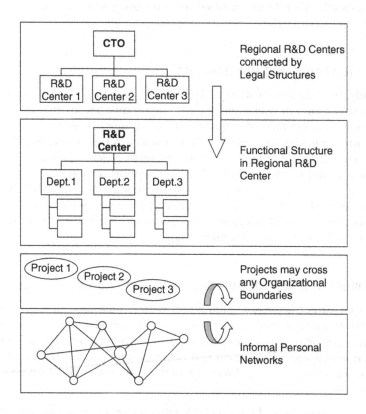

FIGURE 5.3 Four levels of organization in global R&D organizations.

next level brings together all persons who share projects and processes, followed by the more traditional level of a hierarchical and functional structure. As is more common with large global companies, having R&D centers in various countries, a regional and legal structure is set up to manage these labs from one place.

Case Study. Continuous Change and Improvement in Global R&D Management at ABB

The objective of this case study is to use R&D management at ABB as a basis for a broader discussion on various topics, including organization, the search for more innovative projects, different ways of calculating the business impact of R&D projects, and strategic cooperation with universities.

ABB was established in 1988 by the merger of a Swiss company, Brown Boveri, and a Swedish one, Asea. The two companies had very different cultures in their research centers: Brown Boveri focused on basic research, Asea more on applied research. Furthermore, within Brown Boveri, there was some competition between the Swiss and the German research centers. The new chief technology officer (CTO) thus faced a new challenge, for which no model solutions existed: how to create a new culture by combining the best of two worlds.

Figure 5.4 shows the importance of continuously reviewing the situation and coming up with incremental improvements, without destroying what has been accomplished already.

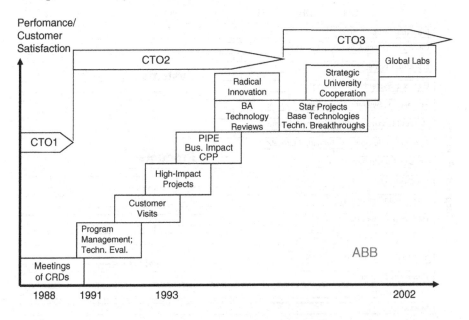

FIGURE 5.4 Continuous improvement in managing R&D (Abbreviations: CTO, chief technology officer; CRD, corporate research director; CPP, cross-program projects; BA, business area).

Meetings of Corporate Research Directors

The initial organization was a typical *line organization*. All research directors reported to the CTO. In joint meetings, it was attempted to come up with compromises about who should do what. It soon turned out that the matter was too complex, there was no room for win/win solutions, and competition rather than cooperation was most often the result.

Program Management

An important step forward was initiated by the next CTO, who established a *matrix organization* between research directors and technology or program managers (PMs) (Figure 5.5). The PMs were responsible for strategy, project creation, project execution, and funding. The research directors were responsible for human resources and operational management of the research centers. Obviously, a key to success was the close cooperation between research directors and program managers. After the initial installment of the PM organization, problems still existed, because the already appointed PMs were not taken seriously by the research directors. Only after the position of head of program managers was created, could a mutual partnership with the research directors be established. To secure the overall cooperation, also with the research managers of the business segments, a technology management team (TMT) under the leadership of the CTO met regularly.

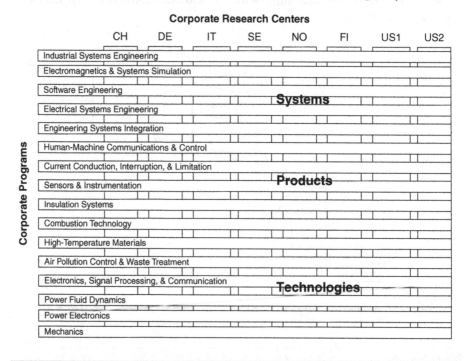

FIGURE 5.5 Matrix organization among 8 corporate research centers and 16 programs.

Once the organization had been established, every year, a few items for improvement were highlighted. The whole organization understood the continuous need to learn and improve, and numerous initiatives were taken to make this happen, such as sharing experiences with world-class professionals outside the company and participating in workshops.

Customer Visits

One example of a highlight, embarrassing as it may seem, was regular customer visits by members of the TMT: meeting with a real customer is much more important than reading about the needs of customers or talking to the sales and marketing organization. It also gives a perspective different from the information obtained by the technical departments of the business units.

High-Risk/High-Impact Projects

Another major step was the introduction of high-risk/high-impact projects, an approach somehow related to internal venturing (or internal entrepreneurship), and discussed earlier.

Cross-Program Projects

One drawback of a technology program was, by definition, its focus on only few technologies. Complex products often are systems containing more than one technology. To become successful in innovation, you have to master the convergence of several technologies. To overcome this barrier and generate projects of larger significance in a systems-oriented corporation, cross-program projects were initiated.

PIPE

Companies have well-established budgeting cycles. New ideas are collected early in the year, discussed with the business units, and proposed as research projects in fall. If approved, these projects receive funding. While this may be a process preferred by accountants, it does not represent reality. One cannot leave idea creation to a short yearly time window. An answer to the request from scientists to *support new ideas*, which came outside the normal budgeting cycle, was presented at one of the regular TMT meetings. To explain the concept to TMT, a graphical model of a pipe was used (Figure 5.6). New projects would move through this pipe all the time. After this concept was understood, the idea was picked up by IT specialists and a Lotus Notes program was developed (see Chapter 6 for more detail) and PIPE now stood for project idea, planning, execution, plus reporting.

Business Impact Calculations

The topic has already been discussed. Simple business impact calculations were introduced early, followed by business-plan-like models for the high-risk/high-impact projects. More advanced and novel methods followed.

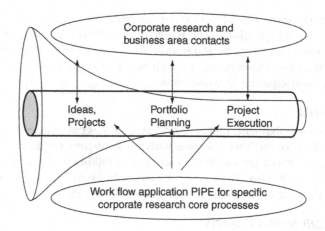

FIGURE 5.6 PIPE as a tool to accept new ideas at any time throughout the year.

Radical Innovation

Radical or disruptive innovations remain an ongoing challenge in a large organization, which quickly settles into its own routines of incremental improvement. This applies equally to product development in a business unit and corporate research. Although the likelihood of success of radical innovations is often low, one has to be aware that it eventually will hit all established technologies. Thus, a focus was given to this issue in corporate research, first by making researchers and business units aware of the issue and establishing regular workshops with the aim of identifying potential areas for radical innovation. Later, this led to the creation of star projects and technology breakthroughs, although with more emphasis on the marketing aspect, "being first and best with a new innovation."

Strategic Cooperation with Universities

All research centers had long ago established cooperation with local universities. Another important improvement, however, was the establishment of strategic cooperation with nonlocal universities. Only a few universities in the United States, the United Kingdom, and Switzerland were selected, and discussions took place at a high management level. It was recognized that very often university research, often because of its basic nature, initiates certain research topics that are of potential interest to the company. Working together with top universities helped provide a look into the future, and it was soon recognized that it also could mean outsourcing of certain research topics. Many academics appreciate the concept of use-inspired basic research, and contact with industry in a strategic partnership gives them inputs about potential needs—a win/win situation.

Virtual Global Labs

As is well-known, a matrix organization also has a disadvantage: Researchers may have two bosses. So, another important step forward was the creation of global labs, which is a virtual organization consisting of several programs. The manager of a global lab can also be the research director of a regional research center, whereas program managers can be simultaneously department managers in a research center.

Concluding Remarks

Managing research, like innovation, does not follow rigid rules. As in research on topics that eventually lead to a new business, one has to continuously be updated about the latest thoughts in management theory. Brown Boveri/ABB corporate research was pretty much an image of the research culture in big companies at that time, although there never was a doubt that it was among the leaders in quality.

Brown Boveri corporate research, after its creation in 1967, was influenced very much by the R&D culture of IBM: technology push. About 10–15 years later, communication with the business units had become more informal, and the importance of market pull was recognized. Then, a phase of contemplation began: Lots of good projects were around, funding was not a problem, but how do we know how good the projects are? There was a continuous search to learn the latest about management thinking on innovation, and through a combination of many events, it was made sure that corporate research was always looking for ways to improve:

- Internal workshops.
- Presentations by some of the leading authors at that time.

The main conclusion was to come up with projects that had large business impact (high-risk/high-impact projects). It was much more challenging to come up with projects that one could call breakthroughs. All those that deserved the name were due to the talent and initiatives of individuals at the medium and lower management levels. What upper management did provide was a very supportive environment. Job rotation and lots of opportunity to have open discussion among R&D, business units, and sometimes even customers were probably the most important factor to facilitate innovation.

SUMMARY

Resistance to change is statistically the right strategy in 95% of all instances, this being equal to the failure rate of new technologies. It is the 5% probability of success of new technologies that is the main driver for continuous change.

In portfolio management the right combination of projects has to be chosen; and with the right attitude toward financing, the emphasis is first on the quality of a new idea, which will automatically bring the right funding.

The barriers to innovation always have to be overcome. Most of these barriers are created by uncertainty, lack of information, and lack of communication. Successful managers learn to recognize them and steer through them relentlessly.

To make innovation a success, commitment at several levels of the organization is important, starting with the champion driving the project, supported by a team, but also having continuous support from top-level management. Since, statistically, failure of projects is frequent, we have to learn to accept failure in innovation and take it as a more important learning experience.

With changes from technology push to market pull to a process continuously involving both aspects, the management of R&D moved from a stage of clear functional separation to a process involving all functions of an organization as early as possible.

The case study of management of R&D at ABB corporate research shows these changes and reflects the continuous search for better organization, more customer contact, and better communication across all levels of the organization. Most important, it shows that the search for more and better innovation is a continuous struggle. The high-risk/high-impact projects were one of several examples how to meet this objective.

REFERENCES

[1] L. Steele, Managing Technology: The Strategic View, McGraw-Hill, New York, 1989.

[2] D.W. Collier, J. Mong, J. Coulin, How Effective is Technological Innovation? Research Management (September–October) (1984) 11–16.

[3] H.R. Zeller, Barriers to innovation in small and medium size companies, presentation at SATW-Transferkolleg, Neuchatel, Switzerland, 2005.

[4] E. Bakke, ABB internal communication, 1998.

[5] O. Lint, E. Pennings, R&D as an option on market introduction, R&D Management 28 (October) (1998) 279–287.

[6] F. Pries, T. Astebra, A. Obeidi, Economic Analysis of R&D Projects: Real Option versus NPV Valuation Revisited, Les Nouvelles (2003) 184–186.

[7] R. Boutellier, O. Gassmann, M. Zedwitz, Managing Global Innovation—Uncovering the Secrets of Future Competitiveness, Springer, Berlin, 2008.

Process Thinking and Knowledge Management

OBJECTIVES

The chapter explains the concepts of processes and work flow, which are important elements of quality improvement. The concepts are applied to the processes important in R&D. The main features of PIPE, a Lotus Notes program used to process the flow of ideas in R&D, are explained.

The Stage-Gate® process is another tool increasingly used in R&D management, introducing a new culture of communication, interaction, and decision making. The objective is to go into some of the details and show its relevance for managing projects across an organization.

The concept of knowledge management received a huge amount of attention after its creation. The objective is to explain the most important features: types of knowledge, the spiral of knowledge creation, and the impact of knowledge management on the types of organizations.

PROCESSES AND WORK FLOW

While the main focus in managing research used to be on the quality of basic research—"as long as we concentrate our efforts on understanding the basics behind some of our core technologies we are sure to find something useful" [1]—additional parameters have come into the picture, the most important one being time.

Craig Tedmon, former CTO of ABB, brought it to the point: "The *time* between when the project is started and when it is finished contains '*future*,' and the future always contains unexpected events, most of which are not helpful to the project. Of course, it is always important to finish a project as fast as possible in order to realize the benefits of the result as soon as possible. Most people understand that. What often is not understood is the former point,

123

namely, the value of shortening the length of the project (*minimizing the amount of 'future'*) in order to minimize unpleasant surprises" (personal communication).

This paradigm shift to consider speed to be the most important factor in global competition has changed the focus from administration of an organization to results and the processes creating results. The IT revolution has made it easier to work across borders in virtual teams. As a result the term *process thinking* has become more important.

Let us begin with a brief definition of processes and work flow.

Processes

From an engineering perspective, industrial processes relate to the sequence of operations, taking up time, space, expertise, or other resources, that lead to the production of some outcome. They may create changes in the properties of one or more objects under their influence.

Examples of processes include

- The process of measurement is the fundamental concept in physics and related sciences.
- The process of materials manufacturing requires equipment, which has to be set up in a workshop and may need modification, is operated by an experienced worker, and takes a certain amount of time to operate.

Work Flow

Work flow can usually be described using flow diagrams, showing directed flows between processing steps. Single processing steps or components of work flow can basically be defined by three parameters:

1. **Input description**. The information, material, and energy required to complete the step.

2. **Steps**. Transformation rules, which may be carried out by human labor, machines, or a combination.

3. **Output description**. The information, material, and energy produced by the step and provided as input to downstream steps.

Work flow components can be plugged together only if the output of one previous (set of) component(s) is equal to the mandatory input requirements of the following component.

By looking at all possible sets of work flow components and modeling them as a work stream path, the efficiency of the flow route through the functions of an organization can be evaluated. Low-efficiency work flows can then be eliminated.

Case Study. ABB: Process Thinking in R&D, from Idea to Global Projects (PIPE and the Stage-Gate® Process)

When the idea of PIPE first arose, the purpose was to describe the importance of accepting new ideas whenever they came up. This led to the concept of continuous program management. To implement the process as effectively as possible, a Lotus Notes software package was developed. To do this, the processes and the work flow had to be identified [2].

Processes and Work Flow

The core processes are the *creation of ideas* (for use within the company) and *execution of projects,* which might lead to new products, prototypes, methods, and basic knowledge.

Process Steps

In research, the *input* to the core processes are ideas, problems, opportunities, brain power, instruments, methods, skills, and much more.

The *steps* might include thinking, experiments, discussions, computer simulations, calculations, manufacturing of prototypes, business negotiations, decisions, reporting, and administration.

The *output*, the "products" of research, consists of ideas, reports, prototypes, methods, articles, presentations, and so on.

There is a clear distinction between *core processes* and *supporting processes.* Typical core processes are research and development, production, and sales. Typical supporting processes are human resource development, strategy development, and budgeting.

Work Flow

The work flow is the horizontal view of a company. It consists of the processes, resources (e.g., personnel, tools, computer systems, money, and suppliers), management, and software applications that support the work flow. Work flow often crosses department and even national borders.

The Challenge

There are many challenges to a modern research organization. Success is due not only to having the best ideas or knowledge. Areas with the highest possible business impact must be highlighted and receive the proper attention, in terms of human and financial resources. Ideas must be communicated and refined in order to realize them. Project teams covering the appropriate areas of competence must be formed.

Important points to be considered to realize the internal challenge include

- Focus on the creation of new ideas and execution of projects.
- Faster process from idea to result.

- Reuse of ideas and information.
- Creation of cross-border project teams and communication support.
- Transparency and less administration.

Continuous Program Management—The Solution

The solution originated from the model of a pipe, which could pick up a good new idea, independent of the time when it was created, and process it by pushing it through the pipe. This was a step forward and away from the traditional process, which asked for the creation of new ideas to satisfy the time schedule of the budgeting process ("invent in spring, refine in summer, present in fall"). Thus, the word *continuous* is important in continuous program management (CPM). New ideas and projects will continuously flow through the system. Decisions about selecting and supporting projects must be taken whenever a good idea or proposal is created, independent of the yearly budgeting process.

CPM binds together the core and supporting processes in a total work flow. To work, well-defined processes and work flows, application tools, computer networks, roles for people in the processes, and databases are required. Cornerstones of CPM are flexibility, speed, simplicity, less bureaucracy, and focus on core processes.

CPM is assisted by a work flow software tool, consisting of applications, databases, interfaces with other software, and computer network functions. Data management makes the system available to everyone who needs it, around the globe, including local accounting systems.

The Tool—PIPE

The work flow tool PIPE (project idea, planning, and execution) consists of three parts with a single user front end: idea creation-planning-project execution.

Idea creation and project execution support the two main processes. The planning application connects these two parts. In planning, resources, strategy, and project proposals are matched (Figure 6.1).

1. **Idea creation.** Everyone within the research community can enter his or her ideas, either using the Lotus Notes platform or directly into the PIPE. Several steps define this process:

   □ **Idea.** At this stage, only the core of the idea, including the potential business impact, is presented. The idea creator is looking for comments to help refine the idea. Ideas can come from anywhere, from inside or outside the company, from patent searches, even "theft" of nonexploited inventions.

   □ **Supported idea.** The aim of this step is to move the idea from the idea stage to a project proposal. Once the idea is supported, more details are needed, such as defined goals, risks, obstacles, and required resources.

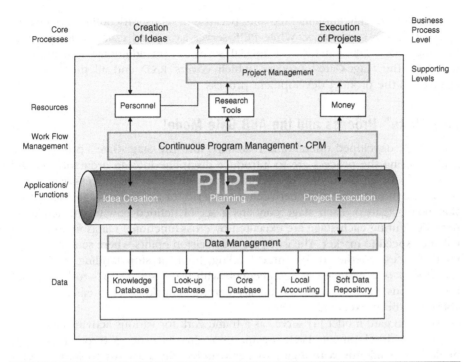

FIGURE 6.1 PIPE: project idea, planning, and execution.

- □ **Project proposal**. Here, the responsible program manager can make the idea a real project proposal (idea acceptance) and transfer it to the PIPE *planning application*.
- □ **Idea bank**. Deferred ideas are saved in the idea data bank for future reference information.

2. **Planning**. The planning application focuses on matching project proposals with customers needs, financial resources, and program strategies. A project leader is assigned and is now responsible for pushing the proposal through *planning*. Large projects can be subdivided into several local projects. The project is then reviewed by the program manager, the local research director, and a business area representative, placing the project into one of three priority groups (ranking). If properly ranked, the project leader asks for local funding and transfers the project into *project execution*.

3. **Execution**. Project execution is a tool for supporting the project leader in his or her day-to-day project management. Project status is summarized by three key elements: status of cost, time, and results, and financial details are entered into the system from the local accounting systems. A report archive function saves the history of the project.

PIPE was quickly implemented and, because of its user-friendliness, accepted by the research community. While PIPE serves as a good example for process thinking and implementation of special software, an additional dimension is added by the Stage-Gate® process, which covers R&D and all the functions involved in the product development process.

Stage-Gate® Process and the ABB Gate Model

Cooper [3] developed operational roadmaps into the Stage-Gate® process. The goals of using the process are to introduce discipline into an ordinarily chaotic process, focus on the quality of execution, and speed up the process. Figure 6.2 shows the model. The steps between are a series of activities (stages) and decision points (gates). Each stage contains a set of defined concurrent activities. Activities during each stage are executed by cross-functional teams in parallel to enhance speed to market. The gates are the decision points where senior management decides whether to continue funding, hold, or stop funding. A team of senior leaders is assigned as gatekeepers, who use a set of rules to help make the decisions. Gates also act as "quality control" checkpoints to ensure deliverables have been executed properly.

The ABB gate model [4] serves as a framework for various activities included in product development. Each organization runs its process through the gate model independently with a gate owner who has the authority to start or stop, fund, and staff the project, including the gate assessor (person who performs assessments on behalf of the gate owner), who is aided by the project leader. Even though no single organization maintains and runs the gate model, ABB created a groupwide process organization to take ownership of improvement and implementation across the company.

The model helped establish a common terminology across the global organization and helped to reduce the number of project ideas (Figure 6.3) in a controlled way, from 300 to 100 within three years. Figure 6.4 describes the ABB Gate Model in its use and application in corporate research.

The gate titles are as follows:

G0, **Start project**. This gate initiates the feasibility evaluation. The focus is on the analysis of the requirements.

G1, **Start planning**. This gate defines the scope of the project. The requirements agreed on here control the planning.

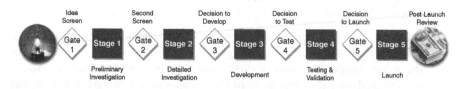

FIGURE 6.2 Cooper's Stage-Gate® model [3].

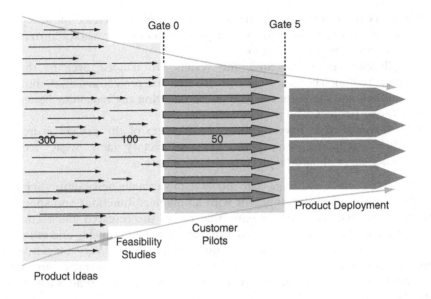

FIGURE 6.3 Funnel development with the gate model.

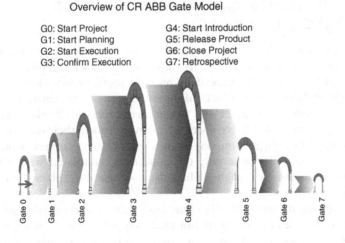

FIGURE 6.4 The ABB gate model.

G2, **Start execution**. This gate marks the agreement on requirements, concept, and project plan. The focus is on specification of functions and architecture.

G3, **Confirm execution**. Here, a confirmation is sought that target dates can be met and the project executes according to the project description and plan. After this gate, the focus is on implementation.

G4, **Start introduction**. The project is now ready for release for acceptance testing. Focus is on validation on preparation for the market introduction and production preparation.

G5, **Release product**. Now, the results are handed over to the line organization. The gate indicates also that the project activities should be finished and therefore the focus is on finalizing any remaining issues.

G6, **Close project**. The project ends.

G7, **Retrospective investigation**. A follow-up of the project is made to check whether the results are satisfactory and what feedback experiences should be given to the organization.

The introduction of a Stage-Gate model led to a totally new culture in executing R&D and product development with all involved functions and speed. If a business unit manager is the engine that drives the research and technology train, the Stage-Gate model is the fuel to make it happen quickly and efficiently. Of course, we have to consider potential disadvantages. Very quickly, if not managed properly, such models can result in new bureaucracies and create new barriers to innovation. Therefore, the main driver behind such models must be speed.

KNOWLEDGE MANAGEMENT, A NEW ASPECT OF ORGANIZATIONAL THINKING

Drucker [5] predicted that typical large businesses would see dramatic reductions in the levels of management and the number of managers; the work would be done more and more by specialists brought together in task forces that cut across traditional departments. Behind these changes lies information technology. Computers communicate faster and better than layers of middle management, and they require knowledgeable users who can transform their data into information. He compares a large company with a symphony orchestra, which has no middle management. There is only the conductor, compared with a CEO, and each of the musicians is a high-grade artist specialist and plays directly to the conductor.

Knowledge management has become one of those terms in management literature that receives all the attention but later may fade away. Definitions of *information* and *knowledge* vary, and you may come to the conclusion that they are used interchangeably. To keep it as simple and understandable as possible, we limit ourselves to a minimum of references [6,7].

Information is data endowed with relevance and purpose. To convert data into information requires knowledge, and knowledge, by definition, is specialized. Information therefore is a set of data organized into a coherent and reusable form (for example, a map). Only data that can be understood can be information. Knowledge is like a trustworthy guide. The guide does not need to consult a map, takes into account recent experience, and can relate your ability to his or her knowledge of the terrain. The guide is the fastest way to achieve your objective, provided you trust that person.

Definition of Knowledge Management

Knowledge management is the identification, optimization, and active management of intellectual assets, either in the form of explicit knowledge held in artifacts or tacit knowledge possessed by individuals or communities.

Explicit knowledge is reusable in a consistent and repeatable manner. It may be stored as a written procedure in a manual or as a process in a computer system, and it can be easily communicated to others.

Tacit knowledge is something we simply know, possibly without the ability to explain: "We can know more than we can tell" (M. Polanyi [8]).

We can see it in artisans or good sporting teams. Human beings are the storage medium; when stored in a community, the danger of loss is reduced, the ability to reuse is enhanced. The act of sharing tacit knowledge always creates something new. Explicit knowledge can be purchased; tacit knowledge is unique, the engine of innovation, and capable of real-time reactivity in decision making. The aim of many companies is to eliminate the need to rely on "experienced characters"; for example, in the steelworks, to ensure consistency of products using sensors rather than relying on experience (tacit knowledge).

Active management of intellectual assets is the creation of management processes and infrastructure to *bring together artifacts and communities in a sustainable form.*

There are four ways to spread knowledge (Nonaka and Takeuchi [9]), also shown in Figure 6.5.

- **Socialization**. Moving tacit knowledge from individuals to communities. This usually starts with building a network of interactions, which facilitates sharing experiences.

	To Tacit Knowledge	*To* Explicit Knowledge
From Tacit Knowledge	Socialization	Externalization
From Explicit Knowledge	Internalization	Combination

FIGURE 6.5 Four modes of knowledge conversion.

- **Externalization**. Changing tacit knowledge to make it explicit. This is triggered by successive rounds of dialogue, in which metaphors and other descriptive examples are used to enable team members to articulate their own views and thereby reveal tacit knowledge that otherwise would not become known.
- **Combination**. Moving explicit knowledge from individuals to communities This is facilitated by coordination among team members, divisions, and business units.
- **Internalization**. Moving from use of explicit knowledge to use of tacit knowledge. This is triggered by the processes of trial and error and learning by doing, which requires articulation of concepts and their concretization.

To realize the practical benefits of tacit knowledge, Nonaka and Takeuchi suggest that it should not only be externalized but also amplified through dynamic interaction among all four modes of conversion. This is also called the *spiral of knowledge creation*, shown in a more understandable modification in Figure 6.6. It implies that the interactions between tacit and explicit knowledge become larger in scale and faster in the course of involving more people, both within and beyond the organization.

There is another dimension to consider as we move up the spiral of knowledge creation: moving up from the level of *nonexpert* to the *expert level*. As we make sense of the senseless (tacit or nonexpert), a group of interdependent individuals who pioneered that understanding develop an expert language that is not yet fully articulated. It is very tacit, based on the common experience of sense making. Over time, this will be codified at an appropriate level of abstraction and pass into a broader community of trainable experts (expert or explicit).

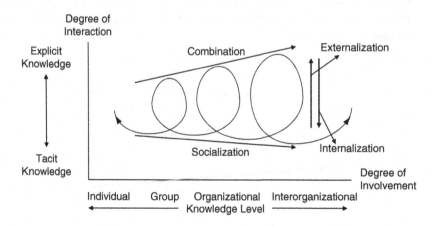

FIGURE 6.6 The spiral of knowledge creation.

This rather complicated excursion away from "common sense" still has the benefit that it allows you to see the importance of these two types of experts and proceed to increase the knowledge level.

Three Faces of Organizations

There are three main images of an organization, depending on the combination of explicit and tacit knowledge: the mechanical machinery, the complex organism, and the dynamic network. All were products of their time, every organization includes elements of all three types.

1. **The mechanical machinery.** Work in an organization is divided and organized in the most effective and profitable way. Operations are directed from one place. According to this early school of thought, the most important characteristics of an organization were predictability, continuity, and manageability.

2. **The complex organism.** Mechanical features are not enough to describe all the phenomena in a real organization. Environment became more important to competition and change. Due to the higher level of education, firms came to realize their employees' skills as a resource for business. In this world, the organization is more like an interactive structure, reality is a living, complex system like a living organism, which easily adapts to changes internally and externally.

3. **The dynamic network.** The third dimension of an organization, the dynamic network, is a response to today's fast-changing environment. Organizations can be viewed as chaotic entities, full of unpredictability and thus unmanageability. The company uses the ability of chaos to organize itself. It manages the fast changes together with its close contacts: interest groups, customers, vendors, and other partners. Chaos functions as the source of innovation—and no other sources exist.

Creative, Competent Workforce

We can now think about ways of how to make the workforce more competent. The term *intellectual capital* had been coined and used mainly in different theories of management and economics. As it does not show up in accounting papers and, to me, somewhat dehumanizes innovative and competent people, I do not use it.

Having looked at the various types of organizations, we can see their different values better if we consider the potential of improving the total combined knowledge base by encouraging continuous information flow through interactions among people with certain competencies. Any researcher who presents a paper at a conference or meeting with other competent people returns with the feeling of increased competence and an awareness of new combinations of thought that

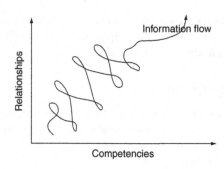

FIGURE 6.7 Information flow through relationships and competencies.

would have been missed had he or she remained isolated in the lab. Figure 6.7 shows how, through information exchange, the knowledge level increases. Information exchange always happens in situations that arise through a combination of relationships among people and their competence.

We can apply this concept to the three types of organization and see that, in a mechanistic command structure with very little human interaction, we make good use of existing know-how, but generate very little new knowledge, whereas organic and dynamic networked organizations have lots of room for new idea creation and continuous improvement (Figure 6.8).

Each type of organization can be characterized by the environment it creates and in which it operates.

Mechanical Corporate Environment

The director is clearly the *visible authority*, in direct use of all the power defined by the following characteristics:

- Personal reliability.
- Clear visions and strategies.
- Good decision-making skills.
- Reporting systems defined by the line organization.

Organic Corporate Environment

The director is now the *initiator of development*, who delegates most power to lower-level management. The management style can be defined by these characteristics:

- Quality management.
- Interpretation forums.
- Feedback systems.
- Performance management discussions.
- Management procedures of processes and teams.

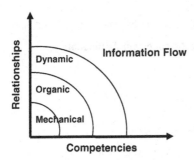

	Mechanical	Organic	Dynamic
Objective	Permanence	Managed growth	Continuous innovation
Knowledge	Defined, explicit	Experiential, hidden, tacit	Intuitive, potential, tacit
Relations	Determined by the organizational hierarchy	Reciprocal, seeking consensus	Spontaneous, networked
Information Flow	One way	Two way	Chaotic
Management Tool	Orders from management	Dialogue, agreed-on working methods, self-assessment	Networking skills

FIGURE 6.8 Competence improvement potential in organizations with more information flow.

Dynamic Corporate Environment

The director is the *visionary*, who focuses on

- Resource allocation.
- Managing innovation.
- Managing partner relationships.
- Portfolio management.

It is easy to see that the trend from mechanical toward the dynamic corporate environment, also facilitated by the advances in information technology, has brought huge advances in successfully running large enterprises.

Table 6.1 lists some regularly cited companies and the aspect of knowledge management they seem to have figured out and for which they are cited.

Although we can argue about the value of formal knowledge management practices, this short introduction can work as an eye-opener toward the best way in which the workforce can communicate and cooperate and why this brings value. As in management of R&D, knowledge management is a process with a starting point but an open end, since we will never be able to accurately predict the future. The ultimate goal of a knowledge management system is to make the whole organization more innovative.

Table 6.1 List of Companies and Their Knowledge Management Applications [10]

Company	Knowledge Management Application
Buckman Laboratories	Knowledge sharing
British Petroleum	Expert networks
Chevron	Best practices database
Dow Chemical	Intellectual asset management
Ernst & Young	Knowledge capture and marketing
Ford Motor Company	"Lessons learned" database
General Electric	Call center
GM Hughes	Reuse of engineering designs
Hewlett-Packard	Sales and marketing support. Knowledge management projects can be found in just about all areas of HP and for a wide range of purposes.
Price Waterhouse Coopers	Best practices database
Sequent Computer	Sales and marketing support
Skandia	Measuring intellectual capital
Teltech	Technical thesaurus

SUMMARY

The concepts of processes and work flow have been explained with the example of the PIPE Lotus Notes tool, which was developed by ABB to manage the innovation process from idea creation to project execution within the R&D function. The tool, open to all members of the research community, has led to the introduction of some kind of standardization in how ideas are recognized and dealt with. Applying the same thinking for all projects, all important questions regarding objectives, planning, ranking, and project execution are addressed.

The Stage-Gate® model, originally developed by Cooper, was introduced in a specialized version at ABB, throughout the whole corporation, including most of the functions. By forcing regulated communication among all players involved in the innovation process, it helped introduce a new culture of thinking and acting and made the process highly efficient. One of the many benefits of this tool is that it helps cut down the number of projects to a few important ones within a short time period.

Knowledge management can be seen as a new and important way to understand how knowledge, which may start out as information stored in our brain, can be transformed into explicit knowledge and how we can assure a fast, efficient way of knowledge creation by continuous interaction among knowledgeable people. The concepts of knowledge management, applied to organizational thinking, helps explain why the move from mechanical organizations to more organic and dynamic ones helps increase the competence of the workforce. Combined with the increased use of IT, this has led to more successful ways to manage a company.

REFERENCES

[1] A.P. Speiser, Brown Boveri internal communication, 1972.

[2] ABB, Internal publication, 1995.

[3] R.G. Cooper, Winning at New Products: Accelerating the Process from Idea to Launch, Perseus Publishing, Cambridge, MA, 2001.

[4] L.P. Chao, K. Ishii, Design Process Error-Proofing: Benchmarking Gate and Phased Review Life-Cycle Models, Proceed. of IDETC/CIE, Long Beach, CA, 2005.

[5] P.F. Drucker, The Coming of the New Organization, Harvard Business Review on Knowledge Management (1998).

[6] G. Bellinger, www.systems-thinking.org/kmgmt/kmgmt.htm.

[7] P. Ståhle, M. Grönroos, Dynamic Intellectual Capital Knowledge Management in Theory and Practice, WSOY, Porvoo, Helsinki, Juva, 2000.

[8] M. Polyani, The Tacit Dimension, Routledge and Kegan Paul, London, 1967.

[9] I. Nonaka, H. Takeuchi, The Knowledge-Creating Company, Oxford University Press, New York, NY, 1995.

[10] F. Nickols, http://home.att.net/~discon/KM/KM_Overview_Exemplars.htm, 2000.

Financial Management of a Company

OBJECTIVES

This very important chapter focuses on various tools we must understand in business. The objectives are to bring you closer to financial accounting—important to know when you run a business or an accountant reports to you. More important in the beginning of your career are the tools of managerial accounting: how to compute costs in production, how to assure profitability with one product or multiple products. The important tool to understand there is marginal costing.

Finally, you will learn what is important in making an investment decision. Here, the special emphasis is on the importance of the net present value method. All these methods are very accurate if applied in hindsight. Looking forward, which is more important, you are faced with risks and uncertainties about the assumptions you make in your calculations.

THE ESSENTIAL INSTRUMENTS TO NAVIGATE THE COURSE OF A COMPANY

To successfully manage a company, some basic workings of the accounting system have to be understood. Long-term success is the result of a series of short-term successes. Depending on the time horizon, different management tools are applicable (Figure 7.1).

Most important in the short term is cash on hand to survive early bankruptcy. The management tools for day-to-day operation and budgeting purposes are the income statement and the balance sheet, whereas future success factors are based on the quality of market research, understanding the needs of customers, and new technologies.

Top managers often compare companies with a large ship, which has to be navigated to keep its course. Measuring instruments support the navigation process, not only on a ship but also in management of a company.

FIGURE 7.1 Short- and long-term success factors in managing a company.

Several books address the subject in an easily understandable way [1–3].

■ **Financial accounting**. In broad terms, accounting provides a financial picture of an organization's activities. Financial accounting is that part which deals with reporting to individuals and groups external to the organization, navigating the company in its external environment. It records all financially relevant connections between the company and its environment. It serves as an account of all the assets, liabilities, ownerships, liquidity, and profitability of the company. Measurements are very accurate but retrospective.

■ **Managerial accounting**. Managerial accounting deals with reporting to individuals inside the organization. It has to serve for the budgeting process and to set up the profit and loss account. It should make it possible to conduct, without too much effort, variance analysis.

■ **Return on investment and capital budgeting**. Several methods exist to calculate the financial consequences of investments.

FINANCIAL ACCOUNTING

Financial accounting is like an information system. It is the process of documenting, measuring, analyzing, interpreting, and communicating information for the pursuit of an organization's goals. It is also used to report to the stakeholders (investors, state, and society). This is also known as *cost accounting*.

The information used in accounting are data from operations of the business, such as turnover, cost, profit, and rate of return. These data are retrospective and describe the actual situation of the company. A problem may arise when they are used for medium and long-term strategic decisions in the enterprise. A positive picture of the past does not automatically lead to a positive picture of the future.

An *engineer* has to learn how to communicate with accountants, because they are responsible for many decisions that affect cost and therefore also the results.

Let us apply this thinking to the example of the surge arrester. Here, the issues of interest are

- Surge arrester technology.
- Design.
- Manufacturing technology.
- Quality assurance.
- Assembly.

Accountants, on the other hand, are responsible for the financial transparency of a company and have to provide quantitative data necessary for the optimum steering of the company. There are three important reporting methods, which, for better understanding, can be shown graphically in a timeline (Figure 7.2).

Balance Sheet: Where We are Now

The balance sheet shows the financial position of a business at one specific time. It shows the financial status at the end of an accounting period. The balance sheet has two sides. The numbers on each side must add up to the same total: the balance sheet balances (Figure 7.3). On one side are the *assets* (things of

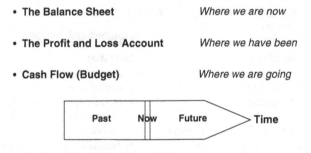

FIGURE 7.2 Accounting in past, present, and future.

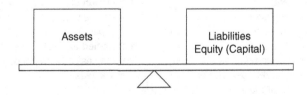

FIGURE 7.3 Graphical representation of the balance sheet.

value the company owns). On the other side are *liabilities* (debts the company owes) and *capital* (the owner's share of the company).

The balance sheet can be described by the equation

$$\text{Assets} = \text{Liabilities} + \text{Capital}$$

A balance sheet is an estimate that can be more or less accurate. The actual list of headings for these three accounts is far more complex. Table 7.1 shows a list of the important headings in a balance sheet.

As an exercise, let us prepare a balance sheet for the Varistor Company using the input data in Table 7.2. Try to consolidate the data into a balance sheet.

Table 7.3 shows the correct way to consolidate the input data into a balance sheet.

Profit and Loss Account (Income Statement): Where We Have Been

The profit and loss account summarizes the results of a company's operations over a period of time (mostly, one year). It is also called the *income statement, operating statement*, and *statement of earnings*. It is the financial statement that shows the bottom line—net profit after taxes. This is the financial report most often used by managers in business. The balance sheet is almost never referred

Table 7.1 List of Important Headings of a Balance Sheet

Assets	Liabilities and Capital
■ **Cash** (Cash available in the bank, or elsewhere to spend) ■ **Accounts receivable** (amounts owed to company by its customers) ■ **Inventory** 　□ **Raw materials** (the stock of materials waiting to be made into products) 　□ **Finished goods** (the stock of completed products ready to sell) ■ **Prepaid expenses** (a payment made in advance, e.g. rent) ■ **Fixed assets** (machinery, equipment, buildings, etc., used over a long period of time in the business.) ■ **Depreciation** (the portion of the original cost of the fixed assets used up or expensed since purchase)	**Liabilities** ■ **Accounts payable** (amounts owed by company to its suppliers) ■ **Notes payable** (amounts owed by company to the bank or other lenders; due within a year or less) ■ **Accruals** (salaries and taxes owed by company but not yet paid) **Capital** ■ **Common stock** (the amount put in by investors to buy the common stock of company) ■ **Retained earnings** (the net profits after taxes, less any dividends paid to stockholders)

Table 7.2 Income Data of Varistor Company (in $ thousands)

Common stock	10,000
Notes payable	125
Shareholding in subsidiary	1,500
Fixed assets	242
Long-term loan	280
Accounts receivable	920
Dividends	390
Vehicles	75
Retained earnings	10
Real estate	16,550
Inventory	1,830
Mortgage	9,385
Cash	196
Accounts payable	560
Reserves	567
Provision for outstanding losses	34
Securities	120
Advance payments from customers	82

to, but the income statement flows from the balance sheet (Figure 7.4). Table 7.4 shows the important headings of an income statement.

As an exercise, let us rearrange the unstructured income statement of the Surge Arrester Company (Table 7.5) in such a way that it shows

- Operating earnings after depreciation.
- Income before taxes.
- Net income.

The various terms used are explained:

- *Sales* is the sum of all prices of the products sold.
- *Cost of materials* is the sum of all prices of materials purchased for manufacturing the products.

Table 7.3 Balance Sheet of Varistor Company

Assets		Liabilities and Equity	
Current assets		Borrowed capital	
Cash	196	Short-term borrowed capital	
Accounts receivable	920	Notes payable	125
Inventory	1,830	Accounts payable	560
		Dividends	390
		Provisions	34
		Advance payments	84
Securities	120	Long-term borrowed capital	
		Long-term loan	280
		Mortgage	9,385
Invested capital		Equity	
Shareholding	1,500	Common stock	10,000
Fixed assets	242	Retained earnings	10
Vehicles	75	Reserves	567
Real estate	16,550		
Balance sheet total	21,433	Balance sheet total	21,433

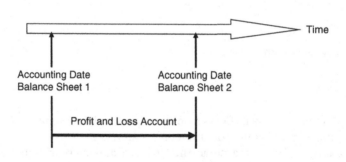

FIGURE 7.4 Evolution of the profit and loss account from balance sheets.

- *Personnel costs* consist of salaries and social benefits.
- *Depreciation* is an expense that reflects the loss in value of a fixed asset (invested capital) and parts of the working capital.
- *Accounts receivable lost* are unpaid invoices of customers.
- *Further general expenses* involve the expenditures for energy, maintenance, repair, insurance, and the like.

Table 7.4 Headings of an Income Statement (separated according to type)

Costs	Revenues

Cost of materials
  Raw materials
  Semifinished products
	Sales
	  Of products

Personnel costs
  Salaries, social benefits
	Interest earned
	  From securities etc.

Interest costs
  Interest on capital; rental cost
	Income
	  From shares

Write-offs (depreciation) of
  Current assets
  Invested capital
	Inventory changes of
	  Raw materials
	  Semifinished products

Taxes On income and capital	
Profit = Sum of the income statement	**Loss** = Sum of the income statement

Table 7.5 Input Data for the Income Statement of the Surge Arrester Company

Costs		Revenues	
Cost of materials	329,515	Sales	840,360
Personnel costs	174,100	Earnings from securities	2,250
Depreciation	77,680	Nonoperating income	830
Accounts receivable lost	7,310		
Further general expenses	197,820		
Nonoperating costs	915		
Provision for taxes	17,200		
Net income	38,900		
Sum	843,440	Sum	843,440

- *Nonoperating costs* are costs independent of the actual business, such as interest expenses.
- *Nonoperating income* is income independent of the actual business, such as value adjustments.
- *Operating earnings after depreciation (net profit before taxes)* are the sales of the firm less costs, such as wages, rent, fuel, raw materials, interest on loans, and depreciation.
- *Net income* is the profit after deduction of taxes.

Cost is the value of money for labor and materials used to produce something. For example, the production of 100 surge arresters requires

2 manufacturing hours	→ cost of salaries
5 machine operating hours	→ cost to operate equipment (depreciation, electric power, cost of auxiliary material)
200 kg raw material	→ cost of various raw materials
storage	→ cost of storage space, interest expenses
production planning	→ cost of labor and plant
transport	→ cost of packaging and freight
accounting	→ cost of labor and office space
advertising	→ cost of labor, office space, and advertising
Miscellaneous	→ further costs

Revenue is the money received from the sale of a firm's products. Let us take as an example the sale of 100 surge arresters. The 100 arresters have been sold for $10,000. This money is called the *income* or *revenue*.

The solution for this exercise is shown in Table 7.6, showing the income statement of the Surge Arrester Company. This multilevel process to establish the profit and loss account permits the subdivision into useful partial results.

What We Can Learn from Balance Sheet and Income Statement

Analyzing financial information serves several purposes:

- For *investors*, analyzing financial information is perhaps the most important element in the financial analysis process.
- For *management*, it provides relevant information about liquidity, profitability, and operational performance.

Table 7.6 Income Statement of the Surge Arrester Company

Sales	**840,360**
Cost of materials	–329,515
Personnel costs	–174,100
Write-offs (depreciation)	–77,680
Accounts receivable lost	–7,310
Further general expenses	–197,820
1. Operating earnings after depreciation	**53,935**
Income from securities	2,250
Nonoperating income	830
Nonoperating expenses	–915
2. Income before taxes	**56,100**
Provision for taxes	–17,200
3. Net income	**38,900**

- For *engineers,* it serves as a source of valuable information about the course taken by the company and that of competing companies.

Information Available by Studying the Annual Report

Annual reports contain a lot of valuable information, but they are usually very optimistic, because the most important readers are the investors. Here are some things you can learn from these reports:

1. It is important to read them year-to-year and compare projected results with achieved results.

2. The graphs shown may slightly distort the truth in a positive direction.

3. Every year one may come to a certain opinion about the future course of the company. It is important to review one's own prediction after one year and look for unexpected changes.

A careful analysis of the various annual reports gives a fairly clear picture of the strategies, products, success in the market, role of government funding, and many more insights. While most of the statements are very optimistic and positive, the section on risks takes mainly a conservative view of what may go wrong. It is as easy to "prove" that the company will go bankrupt within a few years as it is to prove that this is another example of successful technology breakthroughs.

Some annual reports give very good qualitative information about the product range, the competitors (listed by name), major risks, the main ongoing projects and projects in the near future, and offer reasonably accurate figures about next year's income and profitability.

Annual reports are usually very lengthy and contain a lot of information the company is obliged to put in to satisfy the law. Well-written reports make it clear that the major audience is the shareholders, who want to understand clearly where the company has been and where it expects to go. The reader and investor are supposed to walk away with a trust in this information. Whereas technical reports and scientific publications provide an accurate description of the technical situation, a good annual report views the situation from all potential angles: technology, business partners, markets, cash flow, risks, major opportunities, management team.

Financial Ratios

Financial ratios can be classified according to the information they provide. Dozens of financial ratios are available to analyze financial information. The following six types of financial ratios are frequently used: liquidity measurement ratios, profitability indicator ratios, debt ratios, operating performance ratios, cash-flow indicator ratios, and investment valuation ratios.

Liquidity Measurement Ratios

Liquidity ratios are used to measure a company's ability to pay off its short-term debt obligations. This is done by comparing its most liquid assets and its short-term liabilities. The greater the coverage of liquid assets to short-term liabilities, the better, as it shows clearly that a company can pay its debts in the near future and still fund its ongoing operations. Low coverage, on the other hand, is a warning signal for investors that the company may have problems meeting its obligations and running its operations.

The current ratio is the ratio of current assets to current liabilities:

$$Current\ ratio = Current\ assets/Current\ liabilities$$

Profitability Indicator Ratios

Profitability indicator ratios provide a good understanding of how well a company uses its resources in generating profit and shareholder value. The long-term profitability of a company is important both for its survival and the benefit received by shareholders.

Return on investment (ROI) measures the profit a company generates from the resources it has to work with. ROI can be used by management to manage both profits (derived from the income statement) and assets and liabilities (shown in the balance sheet). Of several ROI measures, two are selected.

Return on equity (ROE) shows the interest rate of the owner's capital made available to the company. ROE is the bottom-line measure for the shareholders, measuring the profits earned for each dollar invested in the company's stock:

$$Return\ on\ equity = Net\ profit\ after\ taxes/Owner's\ equity$$

Return on invested capital (ROIC) quantifies how well a company generates cash flow relative to the capital it invested in its business. It is defined as net operating profit less adjusted taxes divided by invested capital, usually expressed as a percentage. When the return on capital is greater than the cost of capital (usually measured as the weighted-average cost of capital), the company is creating value; when it is less than the cost of capital, value is destroyed.

$$Return\ on\ invested\ capital = (Net\ operating\ profit - Taxes)/Total\ capital$$

ROE and ROIC make it possible to compare the interest rates with those of alternative investments. In theory, we can imagine selling (liquidating) the company and investing the capital in the capital market.

Debt Ratios

Debt ratios are used to determine the overall level of financial risk a company and its shareholders face. The greater the amount of debt held by a company, the greater the financial risk of bankruptcy.

Operating Performance Ratios

Operating performance ratios look at how well a company turns its assets (investments in property, plant and equipment) into revenue as well as how efficiently it converts its sales into cash. Important ratios are the fixed-asset turnover ratio and the sales/revenue per employee ratio. Both ratios describe how well the company uses its fixed assets and employees to generate sales.

Fixed Assets Turnover Ratio = Revenue/Investment in Property, Plant and Equipment

This ratio computes how many times the outlay on the investments in these facilities have been covered by the revenues earned.

Sales Revenue per Employee = Revenue/Average Number of Employees

Both ratios depend heavily on the type of industry—if capital intensive, requiring large investments into plant and equipment; or more service oriented industries, which are more labor intensive.

The operating cycle measure is an indication of a company's ability to convert its inventory into cash.

Operating Cycle = Inventory in Days + Receivables in Days

Thus it is the average time period between buying inventory and receiving cash proceeds from the eventual sale of the inventory.

Investment Valuation Ratios

Investment valuation ratios are a series of ratios that can be used by investors to estimate the attractiveness of a potential or existing investment and get an idea of its valuation. These ratios attempt to simplify an otherwise confusing evaluation process by comparing relevant data that help users gain an estimate of valuation. For example, the most well-known investment valuation ratio is the P/E ratio, which compares the current price of company's shares to the amount of earnings it generates. The purpose of this ratio is to give users a quick idea of how much they are paying for each dollar of earnings. And, with one simplified ratio, you can easily compare the P/E ratio of one company to its competition and the market.

Warning Signals

Most ratios by themselves are not highly meaningful. They should be viewed as indicators, with several of them combined to paint a picture of a company's situation. Financial data contained in a balance sheet give information only at a certain point of time. They can easily be off by coincidence or manipulation of data. To be meaningful, most ratios must be compared to historical values of the same company or ratios of similar companies. Financial ratios are used to analyze the past. They fail to recognize future developments.

Cash Flow and Funds Flow (Budget): Where We Are Going

One of the most useful management tools is the cash-flow budget. Preparing this budget forces the manager to consider many facets of the operation. It pinpoints the flow of cash into and out of the company, and it spotlights potential cash flow problems. Most new or small businesses change size and shape quickly. Each new customer means

- A large percentage increase in sales.
- Increase in raw materials.
- More wages and other expenses.

Generally, these expenses have to be met before the customer pays up. Until the money comes in, the business has to find cash to meet its bills. If cash cannot be found, the business becomes "illiquid" and very often goes broke. The cash-flow budget helps determine the best ways of obtaining cash when it is going to be needed.

Effective and profitable management is much more likely if attention is paid to a cash-flow budget. The balance sheet and income statement cannot be as helpful in forcing the kind of planning that will avoid the embarrassment of running out of cash. A company can do well in sales and profits and still go broke. A cash-flow budget can prevent this unwanted occurrence.

Figure 7.5 charts the flow of cash into, through, and out of a business. A study of this diagram can help to improve the understanding of cash flow.

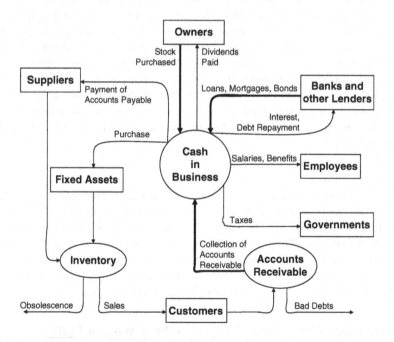

FIGURE 7.5 Diagram showing the flow of cash [1].

MANAGERIAL ACCOUNTING

Financial accounting does not provide all the information needed to plan for the future and steer a company correctly. Therefore, financial accounting has to be supplemented by managerial or cost accounting, which provides sufficient information to manage the company in a target-oriented way.

The objectives of managerial accounting are to permit

- Sufficiently accurate accounting.
- Quick analysis.
- Future-oriented analysis.
- Problem-focused, flexible analysis.

Tools of Managerial Accounting

All tools of managerial accounting have to satisfy the needs of budgeting and profit and loss accounts. In budgeting, the financial goals are defined for the next period as a cooperative effort of all managers with responsibility toward the business.

Which contributions can everyone make to help the company reach its goals? The main input is expected to come from engineers, who are involved in the innovation process, which accounts for more than 90% of the total life cycle of a product:

- Cost of materials (e.g., steel, Al, Cu, Ag).
- Cost of manufacturing (e.g., machining, casting, stamping).
- Cost of logistics (e.g., inventory, just-in-time delivery, transport).
- Installation costs (e.g., assembly, plug and play).
- Maintenance costs (e.g., cleaning).
- Cost of repairs (e.g., repair, replacement).
- Disposal costs (e.g., recycling, hazardous waste).

Costs and Revenues

In financial accounting, expenses and income just represent amounts of money. In managerial accounting, costs and revenues always must be seen in the context with output and are simply the products of volume and unit cost.

It is generally understood that, when the output of a company rises by an amount of 10%, the costs do not rise by the same proportion but by a smaller percentage. On the other hand, when output falls, the percentage reduction in costs is generally less than the percentage fall in costs. The reason for this is the way in which costs are built up and behave as a function of output.

Let us start with a few simple definitions:

- **Costs**. Costs are the values allocated to outputs in goods or services produced by a company:

    Cost = Output volume × Unit cost

- **Revenue**. Revenue is the amount of cash received by a company for the goods or services it provides to customers:

    Revenue = Sales volume × Unit price

- **Profit**. Profit is the amount left when costs are subtracted from the revenues:

    Profit = Sum of revenues − Sum of costs

At first glance the problem looks simple. You just add up the costs and charge a bit more. The more you charge above your costs, provided the customers keep on buying, the more profit you make.

As soon as you start to do the sums, the problem gets a little more complex. For a start, not all costs have the same characteristics. Some costs are independent of how much you sell. If you run a plant, the rent and rates for machinery are relatively constant figures, completely independent of the volume of your sales.

On the other hand, the cost of the products produced in the shop is completely dependent on volume. You cannot really add up those two types of cost until you make an assumption about how many units of your product you plan to make and sell. For example

Rent and rates for the shop	$25,000
Cost of 1000 units of product	$10,000
Total cost	$35,000

Based on this assumed volume we can arrive at a cost per unit of product of

    Total costs/Number of units = $35,000/1,000 = $35

Now, provided we sell out all the units above $35, we will always be profitable. But what happens if we do not sell all 1000 units? With a selling price of $45, we could make a profit of $10,000, if we sell all 1000 units. But, if we only sell 500 units, our total revenue drops to $22,500, and we actually loose $12,500. So, at one level of sales, a selling price of $45 is satisfactory, and at another, it is a disaster.

This very simple example shows that all those decisions are intertwined. Costs, sales volume, selling prices, and profits are linked together. To understand the relationship between these factors, we need a model of how they link up. Before we can build up this model, we need some more information on the types of cost.

The Components of Cost

The last example showed that, if the situation was static and predictable, a profit was certain, but if any one component in the equation was not a certainty (in that example, it was volume), then the situation would be quite different.

The way to address the dependence of costs on volume is to distinguish between those costs which are fixed and those that vary with volume.

Fixed Costs

A fixed cost is a cost or an expense that does not change, independent of how much is produced. Typical fixed costs are marketing costs, R&D costs, cost of administration, and cost of infrastructure (rent, depreciation, interest). Costs such as most of those just mentioned are fixed, irrespective of the time scale under consideration. Other costs, such as those of employing people, while theoretically variable in the short term, in practice are fixed. In other words, if sales demand goes down and a business needs fewer people, the costs cannot be shed for several weeks (notice, holiday pay, redundancy; this situation can, however, be totally different in some countries, where immediate layoffs and rehiring are possible).

Also, if the people involved are highly skilled or expensive to recruit and train (or in some other way particularly valuable) and the slow-down in business looks short, it may not be cost effective to reduce those short-run costs in line with falling demand.

So, viewed over a period of weeks and months, labor is a fixed cost. Over a longer period, it may not be fixed; also, subcontracting to an outside supplier, this cost may not be fixed to the same extent. This is the main reason why large companies often slim down by moving many of their services out of the company and reconnect by making them suppliers.

Figure 7.6 shows fixed costs for a given capacity (left) and an extended one (right), due to new equipment or a new plant.

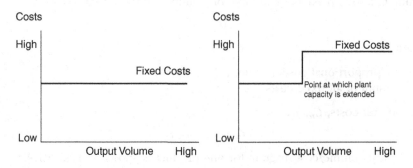

FIGURE 7.6 Fixed costs for a given (left) and an extended (right) capacity.

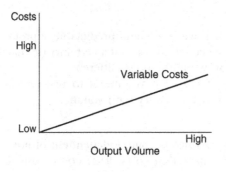

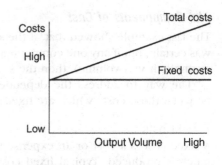

FIGURE 7.7 Linear increase in variable costs as a function of output volume total costs.

FIGURE 7.8 Total costs as a function of output volume.

Proportional or Variable Costs

Proportional or variable costs are all those costs that change as a function of the output (Figure 7.7). Typical variable costs include material costs and labor in a plant. The important characteristic of a variable cost is that it rises or falls in direct proportion to any growth or decline in output volume.

Total costs are the sum of fixed and proportional costs (Figure 7.8). Close observation of the total cost line shows a basic fact: *All increases of total cost are due to increases of variable cost.*

We now enter a territory that can become very difficult: How to determine the product cost for one product—still simple—and how then to obtain the product costs when there are many products—increasingly challenging.

Determining the Product Costs for One Product

Proportional costs can be allocated directly to each individual product, whereas fixed costs are distributed evenly to the total volume of manufactured products. Assuming a total production volume of n units, the total costs TC are

$$\text{TC} = (C_p \times n) + T_f,$$

where

C_p = proportional costs per unit,
T_f = sum of all fixed costs,

and the unit costs, C_u, are

$$C_u = C_p + T_f/n$$

The typical volume/cost diagram for one product is shown in Figure 7.9.

Determining the Product Costs for Several Products

Few companies produce one product, most have a wide range of products, which may even be produced on the same machinery. Also, in this case, the variable

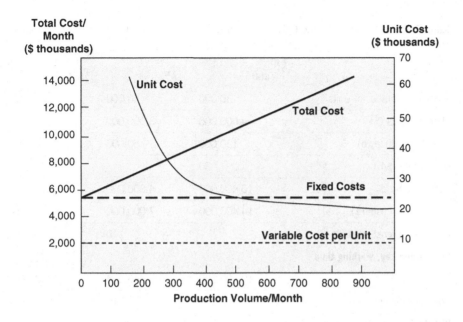

FIGURE 7.9 Unit costs as a function of production volume.

costs can be directly allocated to the various products. The fixed costs, however, have to be distributed among the various products according to a *distribution key*. Typical distribution keys are

- Variable costs.
- Working time.
- Production volume.

For working time, t, as a distribution key, the respective fixed costs for a product, A, are

$$C_{fA} = C_f \times t_A/t_{tot}$$

where

C_{fA} = respective fixed costs for product A.
C_f = the sum of all fixed costs.
t_A = the total working time to manufacture product A.
t_{tot} = the total working time to manufacture all products.

and all data refer to a specific time period (e.g., one year).

For example, the Surge Arrester Company produces two types of surge arresters, Type A5 and Type A10. Type A5 is made of porcelain, which has to be assembled from many single components, whereas Type A10 is a modern polymer-based arrester, consisting of only a few parts.

Table 7.7 The Influence of the Choice of Distribution Key for Fixed Costs on Cost per Unit

	Total	Share A5	Share A10
Produced number of units	30,000	50,000	250,000
Material costs ($)	4,000,000	3,000,000	1,000,000
Time to produce (h)	130,000	50,000	80,000
Hourly rate ($/h)	80		
Cost of labor ($)	10,400,000	4,000,000	6,400,000
Proportional costs ($)	14,400,000	7,000,000	7,400,000
Proportional cost per unit ($)		140	30
Distribution key, working time			
Fixed costs ($)	10,000,000	3,846,154	6,153,846
Cost per unit ($)		217	54
Distribution key, prop. costs			
Fixed costs ($)	10,000,000	4,861,111	5,138,889
Cost per unit ($)		237	50

We assume a time period of one year for the production of arresters. As distribution key for the fixed costs, we use 1 times working time and 1 times proportional costs (Table 7.7).

Cost Accounting Systems

The purpose of cost accounting is to provide a basis for decision making in an organization.

Full-Cost Accounting

The concept of full-cost accounting is that each product sold above its full cost generates a profit. In full-cost accounting, all costs generated in the company, both variable and fixed costs, are allocated to the products (Figure 7.10). Thus, full-cost accounting is based on the value-added process in a company.

The problem with full-cost accounting is that additional sales volume sold at prices between the variable costs and full costs is "given away." Nevertheless, this business also *contributes* to the profit, because it increases only the total variable costs. Any revenue above the variable costs makes an additional contribution in reducing the fixed costs.

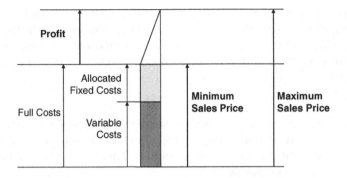

FIGURE 7.10 Concept of full-cost accounting.

For example, the Surge Arrester Company sells

50,000 arresters of type A5.
250,000 arresters of type A10.
The sales price for A5 is $220, and for A10, it is $90.
Total revenue: R = 50,000 × 220 + 250,000 × 90 = $33,500,000

The total costs for the production of arresters using the time-based distribution key are

Type A5 $10,846,154
Type A10 $13,553,846
Total $24,400,000

The profit can be calculated as

$$\text{Profit} = \text{Revenue} - \text{Costs} = 9,100,000$$

The profit contribution (key is t) per part is

$$\text{Type A5} : 220 - 217 = \$3$$
$$\text{Type A10} : 90 - 54 = \$36$$

The profit contribution (key is variable cost) per part is

$$\text{Type A5} : 220 - 237 = -\$17$$
$$\text{Type A10} : 90 - 50 = \$40$$

Depending on the distribution key applied for the fixed costs, a product may appear profitable or unprofitable!

Marginal Costing (Direct Costing)

The idea behind marginal costing is to keep sales prices adjustable to individual market prices in the most flexible way, such that each business can be contracted which makes a positive contribution to the profit of the company. Contrary to full cost accounting only a portion of the total costs, basically the variable costs will be allocated to the products. The remaining costs for the company—the fixed costs— have to be covered by the difference between revenue and the variable costs.

The *unit contribution margin (CM)* is unit revenue or price (*P*) minus unit variable cost (*V*):

$$CM = P - V$$

In other words, the contribution margin is the marginal profit per unit sale.

Starting points in margin costing are therefore not the actually rising costs, but the revenues available under market conditions. This method is used particularly for short-term decision making.

Figure 7.11 shows the situation of revenue coming from three products, each having different variable costs and contribution margins. The most important finding is this: *To make a profit, the sum of all contribution margins has to be larger than the fixed costs*.

The following definitions are quite common and useful to remember:

Gross profit = Revenue − Variable costs = Contribution margin
Gross margin (%) = Gross profit/Revenue × 100

Gross margin is a good indication of how profitable a company is at the most fundamental level. Companies with higher gross margins have more money left

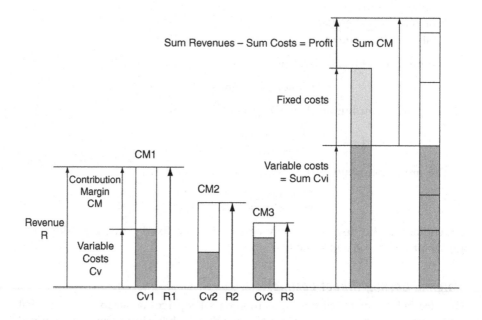

FIGURE 7.11 Concept of marginal costing.

over to spend on other business operations, such as research and development or marketing.

$$\text{Net profit} = \text{Sum of gross profit} - \text{Fixed costs}$$
$$\text{Net margin (\%)} = \text{Net profit/Revenue} \times 100$$

In general, the "profit" in a company is net profit:

$$\text{Net profit} = \text{Sum of revenues} - (\text{Sum of variable} + \text{fixed costs})$$
$$= \text{Sum of revenues} - \text{Sum of costs} = \text{"Profit"}$$

Salespeople often are tempted to sell below full cost to generate more sales. If, however, too many products are sold below full cost but with a positive contribution margin, the sum of the contribution margins can be less than the fixed costs, which leads to losses.

Breakeven Analysis

A common method of presenting cost information to management is in the form of breakeven charts. Although lots of detailed costs lie behind breakeven charts, the graphical presentation shows only the relevant facts necessary for management decisions. It is rarely necessary to know all the "pertinent" cost details.

The breakeven point for a product is the point where total revenue received equals total costs associated with the sale of the product (TR = TC). A breakeven point is typically calculated for businesses to determine if it would be profitable to sell a proposed product or to determine the minimum sales volume for given costs. Breakeven analysis can also be used to analyze the potential profitability of expenditure in a sales-based business. The breakeven point can be defined by either a critical sales volume (monetary) or a critical number of units sold:

$$\text{Breakeven point} = \text{Total fixed costs/(Unit selling price} - \text{Average variable costs)}$$

The price minus the average variable cost is the variable profit per unit, or contribution margin of each unit sold. This relationship is derived from the profit equation:

$$\text{Profit} = \text{Revenues} - \text{Costs}$$

where

$$\text{Revenue} = (\text{Unit selling price} \times \text{Number of units})$$

and

$$\text{Costs} = (\text{Average variable costs} \times \text{Number of units}) + \text{Total fixed costs}$$

Therefore,

$$\text{Profit} = (\text{Unit selling price} \times \text{Number of units})$$
$$-(\text{Average variable costs} \times \text{Number of units} + \text{Total fixed costs})$$

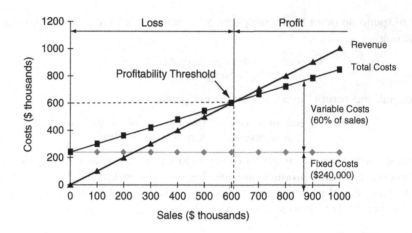

FIGURE 7.12 Determination of breakeven sales volume.

Since profit at break-even equals zero, the number of units of a product at break-even is

$$\text{Total fixed costs} / (\text{Unit selling price} - \text{Average variable costs})$$

Starting from the planned or budgeted fixed costs (e.g., $240,000) and the average contribution margin (e.g., 40% of sales volume), the breakeven sales volume can be determined (Figure 7.12).

We can also read off the amount of profit or loss at any point by looking at the difference between the total revenue and total cost lines.

A similar graph, called a *contribution breakeven chart*, shows sales revenue, total costs, and variable costs (e.g., variable instead of fixed costs). To plot this graph, we first draw in total variable costs and then move the line up by the amount of fixed costs. Figure 7.13 reveals the existence of different levels of output and value, which represent profit, loss, and breakeven. Breakeven is depicted at the junction of the total cost and revenue lines. Profit is depicted by the wedge to the right of the breakeven point, where the revenue line is shown at a higher value than the total cost line, and loss is shown to the left of this point where the total costs are shown at a higher value than the revenues. *Careful observation of the figure shows that the maximum loss shown is equal to the fixed costs.*

Any item of fixed cost, be it the top manager's salary or the rent of the premises, is totally nonproductive in terms of earning profit and must be financed from the revenue generated by the sales of the company's products.

Contribution

We now want to consider why profits occur and, further, why the maximum loss decreases as output increases up to breakeven point and profit occurs beyond this point. Looking at Figure 7.13, we see that the maximum loss decreases

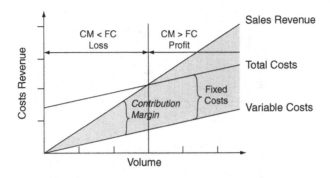

FIGURE 7.13 Contribution breakeven chart.

because of the difference that arises between variable costs and revenue. This difference between total revenue and variable costs, as defined earlier, is called *contribution*, and at the breakeven point, it exactly equals the fixed costs. As the name implies, it represents the contribution that sales make, individually and in total, to the fixed costs of the company. Once the sum of all the contributions is large enough to cover the fixed cost (breakeven), the business starts to make profit, which increases as the output increases. We can sum it up: Profit is the contribution of all units of output above the breakeven point.

Margin of Safety

Once a breakeven chart has been constructed for any given business situation, one of the most important measures is the current level of sales. The difference between this level of output and the breakeven point (Figure 7.14) is known as the *margin of safety*. Management would like to see large margins of safety, because this permits a significant drop in sales before starting to incur losses.

Let us try another exercise. In the previous example about the Surge Arrester Company, we saw that, depending on the distribution key for fixed cost (variable cost or labor), a product, such as arrester type A5, can become unprofitable. We continue to work with the assumptions in the earlier example and assume further only the variant with fixed costs distributed by using variable costs as distribution key.

Questions

1. Would you, as a business unit manager of the arrester company, continue to keep arrester type A5 in your assortment or drop it? Explain your decision.

2. How large is the gross margin for arresters type A5 and type A10?

3. You have the opportunity to receive a large order for additional 50,000 arresters, type A10. The customer, however, does not want to pay more than $1.4 million. Would you accept the order? Explain your decision.

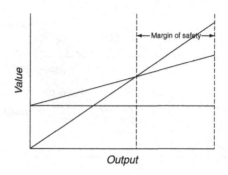

FIGURE 7.14 Margin of safety.

4. Investigate, if arrester type A5 can be profitable assuming
   - Distribution of fixed costs using variable costs as key.
   - Variable costs of $140 per unit.
   - An average sales price of $220 per unit.

Solutions

1. A comparison of full cost accounting with margin costing (Table 7.8) shows that type A5 alone does not give a positive profit, but nevertheless contributes with a positive contribution margin to help cover the fixed costs. Without the A5, profit would shrink from $9.10 to $5.10. Therefore, A5 should be kept in the assortment, but a follower product should be introduced with much lower variable costs to provide a positive contribution. If sales price of A5 drops to $140 (= variable costs), this product would have to be dropped immediately, because it does not provide a positive contribution margin.

2. The gross margin, GM, is the ratio of contribution margin and revenue; thus,

   Type A5: GM = $4 million/$11 million × 100 = 36%
   Type A10: GM = $15.1 million/$22.5 million × 100 = 67%

3. For an order of 50,000 arresters type A10 and a total price of $1.4 million, the revenue per unit (r) is

   $$r = \$1,400,000/50,000 \text{ units} = \$28/\text{unit}$$

   This value is clearly below the variable costs of $30 for type A10. The contribution margin would become negative, so the order should be declined from a cost analysis point of view. To view this from a different aspect, if other products are sold to the same customer, which are profitable, A10 should remain in the portfolio because of strategic marketing considerations.

4. The breakeven analysis for type A5 gives

   Breakeven = Fixed costs × 100/Unit selling price − Unit variable cost
   = $4,861,111 × 100/36 = $13,503,086

Table 7.8 Comparison of Full-Cost Accounting with Margin Costing

	Type A5	Type A10	Total	Total without Type A5
Number of units sold	50.000	250.000	300.000	250.000
Variable costs/unit	140	30		
Revenue/unit	220	90		
Full-Cost Analysis				
Revenue	11,000,000	22,500,000	33,500,000	22,500,000
Variable costs	7,000,000	7,400,000	14,400,000	7,400,000
Fixed costs (prop. VC)	4,861,111	5,138,889	10,000,000	10,000,000
Result	−861,111	9,961,111	9,100,000	5,100,000
Marginal Costing				
Revenue	11,000,000	22,500,000	33,500,000	
Variable costs	7,000,000	7,400,000	14,400,000	
Contribution	4,000,000	15,100,000	14,100,000	
Fixed costs			10,000,000	
Result			9,100,000	

To reach the breakeven point for type A5, the sales volume has to be increased from $11 million to above $13.5 million to make type A5 profitable. Assuming an average sales price of $220, a minimum sales volume of $13,503,086/$220 = 61,378 units. This corresponds to an increase in the production volume by about 23%. This also causes an increase in variable costs of about 23%, which then increases the portion of the fixed costs for type A5 (by about 10%). To compensate for the increase in the fixed costs, an additional increase in the sales volume is needed and so forth. Several iterations are needed to determine the breakeven point with reasonable accuracy. After five iterations, a sales volume of about $15.9 million and 72,000 arresters is determined.

Final Remarks and Key Learning Points

The distinction between fixed and variable costs changes depending on whether the decision is short or long term. Defining fixed and variable costs is to be considered relative, since

- Variable costs are not really proportional to output.
- When sales drop, employees cannot be fired immediately.
- When sales increase suddenly, more expensive overtime is required.
- Higher purchasing volume means lower material costs.

Fixed costs are not always independent of produced volume:

- Larger sales volume may require additional equipment and infrastructure.
- More people in sales and marketing and in administration are needed.

The distribution of fixed costs onto products can be totally wrong in certain cases:

- Selling to a large customer requires lower costs for selling and acquisition.
- Product customization may cause additional development costs.

Current trends are toward higher levels of fixed costs for all types of businesses, due mainly to increased levels of technology investment. This investment is represented by both the annual depreciation charge on the equipment and the cost associated with salaries, maintenance, and higher labor costs, which reflect increasing specialization. The consequence of such a trend is that breakeven points are at a higher level of capacity, with a consequent reduction in the margin of safety at normal levels of activity. The inevitable result of such a situation is that *management's control of operational costs needs to become more accurate*.

One aspect of this control is a clear understanding of the factors that contribute to the profitability of current operations:

- Identification of the variable cost of each product and its contribution at different selling prices.
- A constant analysis of the sales mix of each product and its output level.
- Identification of fixed costs and their constant monitoring to eliminate wasteful expenditures.

The key learning points are

- All costs can be separated into fixed and variable categories.
- All increases in total cost are increases in variable costs, reflecting increases in output.
- The maximum cost that a company can incur is equal to its fixed costs.
- The difference between the revenue and the variable cost is the contribution.
- Profit is contribution on all units of output above breakeven point.

INVESTMENT DECISIONS: CAPITAL BUDGETING

Decisions on investment, which take time to mature, have to be based on the returns that investment will make. If the investment is unprofitable in the long run, it is unwise to invest in it now. Investments are made to quantitatively or qualitatively change the production and sales capacity of a company to achieve higher profits or lower costs. The term *investment* is used differently in cost accounting and in financial accounting. Typically, investment involves using financial resources to purchase a machine, building, or other asset, which then yields returns to an organization over a period of time. While this is how economists in a company refer to a real investment, financial economists refer to a financial asset, such as money that is put into a bank or the market, which may then be used to buy a real asset.

Key considerations in making investment decisions are

1. How much should be invested—can the company afford it?

2. How long will it be before the investment starts to yield returns?

3. How long will it take to pay back the investment?

4. What are the expected profits from the investment, and are there alternative ways to invest the money to secure a higher return?

Examples of investment projects include the purchase of manufacturing equipment, building a whole new plant, purchase of transportation equipment, and buying a patent or a license. Research and development of a new product or service is also an investment. So is acquisition, the investment into another company. To sum it up: An investment is any project that requires a capital expenditure and generates a future cash flow.

Because capital expenditures can be very large and have a significant impact on the performance of a company, great importance is placed on the process of investment appraisal to select the right projects. The purpose of an investment appraisal is the determination of the financial consequences of investment decisions. Therefore, it is focused on individual investment objects and takes into account the total utilization period. This is in contrast to financial and managerial accounting, which deals with the whole company and is limited to a period of one year or less.

Typical characteristics of investments are a utilization period of several years, an orientation toward the future and the resulting benefit, and the need for financing if the magnitude of an investment lies above certain limit.

Referring to the preceding examples, we can distinguish among several types of investment:

- Investments into new equipment, new plants, new vehicles, and the like.
- Financial investments (investing in other companies, acquisitions).
- Immaterial investments (research and development projects, patents, licenses).

These can be further subdivided into investments for replacement, growth, rationalization, and new investments.

Economic Parameters for an Investment Appraisal

The most important parameters for an investment appraisal are

- Cost of investment.
- Economic benefit.

- Utilization period.
- Income from liquidation (resale value).
- Time value of money (calculated interest rate).
- Risk.

Cost of Investment

The investment costs are at the beginning of an investment calculation, and they are the most precise part of the calculation. In the example of investing in production equipment, we can calculate the investment cost as the sum of various positions:

Cost of investment = Net price of equipment + Freight costs + Cost of installation
+ Secondary investments (upgrading buildings, power supply, etc.)
+ Interest rate on prepaid expenses

Economic Benefit

Economic benefit is the benefit resulting from all additional revenues and saved expenses during the lifetime of the investment. Economic benefits can also result from additional sales or higher expenses in one organizational unit but lower expenses in another one.

Utilization Period

The period of utilizing the investment depends on various factors, which contribute to the devaluation of an investment:

- Physical degradation.
- Technological progress.
- Legal requirements.

Income from Liquidation (Resale Value)

The income from liquidation is the resale value at the end of the utilization period minus the cost of removal and disposal.

Time Value of Money

The interaction with lenders with borrowers sets an equilibrium rate of interest. Borrowing is worthwhile only if the return on the loan exceeds the cost of the borrowed funds. Lending is worthwhile only if the return is at least equal to that which can be obtained from alternative opportunities in the same risk class.

The interest rate received by the lender is made up of

- **The time value of money**. The receipt of money is preferred sooner rather than later. Money can be used to earn more money. The earlier the money is received, the greater the potential for increasing wealth.

- **The risk of the capital sum not being repaid**. This uncertainty requires a premium as a hedge against the risk; hence, the return must be commensurate with the risk being undertaken.
- **Inflation**. Money may lose its purchasing power over time. The lender must be compensated for the declining purchasing power of money. A lender who receives no compensation will be worse off when the loan is repaid than at the time of lending the money.

There are two objectives in using this interest rate by the investor:

- Discounting to the net present value allows a comparison of expenses and income originating at different times. Money made in the future is worth less than money made now.
- It shows the capital costs of an investment. These can be compared with either the effective average capital costs of a company or the rate of return of an alternative capital use.

Risk

To allow for unpredictable cost increases due to risk factors, an additional increase in the interest rate is applied.

Methods Used in Capital Budgeting

The various methods used in capital budgeting can be classified in two groups. Depending on disregarding or regarding the time value of money, one can distinguish between the static method and the dynamic method.

In the *static method*, all present and future expenses and revenues are treated equally. No consideration is made of the time value of money.

In the *dynamic method*, all future expenses and revenues are discounted back to the time of the investment (net present value method).

The possibility of earning interest with the passage of time means that, even in a situation without price inflation, all future costs and revenues must be discounted back at an appropriate interest rate before they can be properly compared with expenses occurred or revenues received today. To properly compare revenues and expenses, one has to adjust them to the same time period. In general, this is the present time, because an investor is able then to judge if an investment is worthwhile or not.

Assuming an annual interest rate, k, a capital, C_o, which is invested today, will grow in n years to a future value of

$$C_n = C_o(1 + k/100)^n$$

where

C_o = the initial sum invested (present value).
k = the interest rate.
n = the number of periods for which the investment is to receive interest.

We can derive the present value C_o by using the formula

$$C_o = C_n/(1 + k/100)^n$$

Decision Criteria for Investments

Three methods, out of many more, are chosen as representative: the static method (payback calculation), dynamic methods (net present value calculation), and the internal rate of return.

Payback Calculation

This is the most important *static method* in investment calculations. The payback time is the time needed to recover all investment costs by the sum of all profits (cash flow) due to the investment:

Cash flow per period = Profit or cost savings per period

Two methods exist to determine the payback time:

- **Payback period**. This method asks how long it will take to get back the money that has been invested, assuming a constant value for the cash flow over time. Payback period (or payback time), as a tool of analysis, is often used because it is easy to apply and easy to understand. When used carefully or to compare similar investments, it can be quite useful:

  Payback time = Investment/Cash flow

  If a project pays back its investment in five years, it has a payback period of five years. A five-year payback is better than a ten-year payback, all other things being equal.
- **Cumulated cash flow**. For each period, the cash flow is calculated and cumulated until the sum of all cash flows is equal to the investment.

Let us perform an exercise in payback calculation. In 2000, the Surge Arrester Company had to replace its old sintering furnace, which is used in the production of arresters. Two sintering furnaces with different functionalities have to be evaluated using the data in Table 7.9:

1. Compute the payback time for both furnace types using both an averaging payback-time calculation and the cumulated cash-flow method. Discuss the results.

2. How does the result change if the time value of money is considered in the cumulated cash-flow analysis for furnace B? Discuss the result.

Using payback-time calculation,

Furnace A: Profit = Revenue − Costs = 1,150,000 − (625,000 + 240,000) = \$285,000

Payback time = 1000/285 = 3.5 years

Table 7.9 Comparison of Two Investments

	Furnace A Partially Automated	Furnace B Fully Automated
Investment costs (year 0) ($)	1,000,000	1,600,000
Useful economic life (years)	8	10
Salvage value ($)	0	0
Capacity (tons/year)	500	600
Potential annual sales		
Until year 6 (tons)	500	500
Starting in year 7 (tons)	600	600
Net profit per ton ($)	2,300	2,300
Net annual profit ($)	1,150,000	1,150,000
		1,380,000
Average annual net profit ($)	1,150,000	1,242,000
Cost per ton ($)	1,250	1,100
Annual costs ($)	625,000	550,000
		660,000
Average annual costs ($)	625,000	594,000
Overhead costs/year ($)	240,000	300,000
Desired interest rate %	12	12

$$\text{Furnace B: Profit} = 1,242,000 - (594,000 + 300,000) = \$348,000$$

$$\text{Payback time} = 1600/348 = 4.6 \text{ years}$$

1. Using the payback-time method (only for furnace B), we show the results in Table 7.10 and Figure 7.15. If the cash flow varies over time then the payback method is very inaccurate.

2. For furnace B, we consider the time value of money. Assuming a 5% interest rate, the cumulated cash flow for 2001 (also in thousands of dollars) was

$$C_0 = -1,600 + 300/(1 + 5/100) = -1,314 = C_{01}$$

The cumulated cash flow for 2002 (in thousands of dollars) was

$$C_0 = -1,600 + 300/(1 + 5/100) + 300/(1 + 5/100)^2 = C_{01} + 300/(1 + 5/100)^2 = -1,314 + 272 = -1,314 + 272 = -1,042$$

And, generally, for the year 200n (Table 7.11 and Figure 7.16),

$$C_n = C_{0(n-1)} + C_n/(1 + k/100)^n$$

The payback method is easy to apply and therefore frequently used for the first evaluation of an investment project.

Table 7.10 Payback Time for Furnace B, Using the Payback-Time Method ($ thousands)

	2000	2001	2002	2003	2004	2005	2006	2007	2008
Investment cost	-1600								
Costs		-550	-550	-550	-550	-550	-550	-660	-660
Overhead costs		-300	-300	-300	-300	-300	-300	-300	-300
Net profit		1150	1150	1150	1150	1150	1150	1380	1380
Cash flow/year	-1600	300	300	300	300	300	300	420	420
Cumulated cash flow	-1600	-1300	-1000	-700	-400	-100	200	620	104

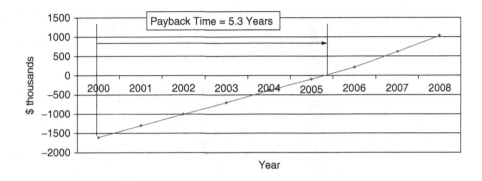

FIGURE 7.15 Cumulated cash flow.

The advantage of this method is that it is a good indicator in judging risk of an investment project: The shorter the payback time, the lower the risk. It has, however, the disadvantage that it gives only limited information about actual rate of return, especially over long periods.

Net Present Value Method

The net present value method is a dynamic method and tries to show whether the future profits, discounted by an interest rate adjusting for time and risk, reach the value of the investment:

$$\text{Cash value of all profits} - \text{Investment sum} = \text{Surplus?}$$

For NPV > 0, we obtain a surplus. An investment is better the higher the surplus. If NPV < 0, the investment subtracts value from the company and should be rejected. Surplus can also be converted into an "interest rate" by relating it to the investment.

The general equation for net present value is

$$\text{NPV} = X_0 + X_1/(1+k) + X_2/(1+k)^2 + \cdots + X_n/(1+k)^n$$

where

$X_0 = $ the initial cash outflow.
$n = $ the final period, in which cash inflows or outflows are received.
$k = $ the required rate of return.

Let us perform an exercise in net present value calculation. We evaluate the furnace project of the Surge Arrester Company using the net present value method.

Furnace A
Investment cost $1,000,000
Period of utilization   8 years
Cash flow per year $285,000

Table 7.11 Payback Time for Furnace B, Using the Cumulated Cash-Flow Method ($ thousands)

	2000	2001	2002	2003	2004	2005	2006	2007	2008
Investment cost	−1600								
Costs		−550	−550	−550	−550	−550	−550	−660	−660
Overhead costs		−300	−300	−300	−300	−300	−300	−300	−300
Net profit		1150	1150	1150	1150	1150	1150	1380	1380
Cash flow/year	−1600	300	300	300	300	300	300	420	420
Cum. CF ($k = 0\%$)	−1600	−1300	−1000	−700	−400	−100	200	620	104
Cum. CF ($k = 5\%$)	−1600	−1314	−1042	−783	−536	−301	−77	221	505
Cum. CF ($k = 10\%$)	−1600	−1327	−1079	−854	−649	−463	−293	−78	118
Cum. CF ($k = 15\%$)	−1600	−1339	−1112	−915	−744	−594	−465	−307	−169

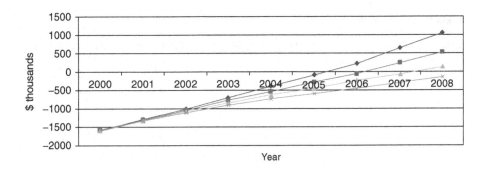

FIGURE 7.16 Discounted cash flow.

Assuming a 12% interest rate a net present value (surplus) can be computed:

$$NPV = 285 \times \sum_{n=1}^{8}(1 + 12/100)^n - 1{,}000 = 285 \times 4{,}968 - 1{,}000 = 416$$

Furnace B
Investment cost $1,600,000
Period of utilization   10 years
Cash flow per year
Year 1–6 $300,000
Year 7–10 $420,000

We assume again a 12% interest rate and can discount the cash values as in Table 7.12. Again, we get a surplus, which means that the internal rate of return is larger than 12%.

If we relate the surplus to the investment we obtain profitability for

Furnace A: 416/1,000 = 41.6%
Furnace B: 280/1,600 = 17.5%

Internal Rate of Return

The internal rate of return (IRR) is a rate of return on an investment. The IRR of an investment is the interest rate that gives it a net present value of 0, or where the sum of discounted cash flow is equal to the investment. The IRR is calculated by trial and error.

IRR is the best method to evaluate the economic side of an investment, because it allows a good comparison with other investment projects and financial alternatives (bank account, stocks, real estate):

$$X_0 = X_1/(1+r) + X_2/(1+r)^2 + \cdots + X_n/(1+r)^n$$

Table 7.12 List of Discounted Cash Values

Year	Discounting Factor	Cash Flow	Discounted Value
1	0.893	300	208
2	0.797	300	239
3	0.712	300	214
4	0.386	300	191
5	0.567	300	170
6	0.507	300	152
7	0.452	420	190
8	0.404	420	170
9	0.361	420	151
10	0.322	420	135
Sum of all discounted values			1880
minus investment			−1600
Surplus (NPV)			280

If the initial cash outflow occurs at time 0, it is represented by the interest rate r.

Let us perform an exercise in the IRR method. We evaluate the furnace project of the Surge Arrester Company using the internal rate of return method.

1. Furnace A provides a yearly constant cash flow:
   - □ Investment sum: $1,000,000
   - □ Utilization period: 8 years
   - □ Mean annual revenue: $1,150,000
   - □ Mean annual costs: $625,000
   - □ Overhead costs: $240,000

   From this a payback time of 3.5 years and an annual net income of $285,000 can be calculated. According to definition, the *internal rate of return* is the interest rate, for which

$$\text{Investment sum} = \text{Discounted cash flow}$$

   In our example, this means

$$1,000 = 285 \times \sum_{n=1}^{8} (1 + k/100)^n$$

$$3.5 = \sum_{n=1}^{8} (1 + k/100)^n$$

Table 7.13 Discounting Factors for Several Interest Rates and Years

Year	10%	15%	16%	20%	22%	23%	24%
1	0.909	0.870	0.862	0.833	0.820	0.813	0.806
2	1.736	1.626	0.605	1.528	1.492	1.474	1.457
3	2.487	2.283	2.246	2.106	2.042	2.011	1.981
4	3.170	2.855	2.798	2.589	2.494	2.448	2.404
5	3.791	3.352	3.274	2.991	2.864	2.803	2.745
6	4.355	**3.784**	**3.785**	3.326	3.167	3.092	3.020
7	4.868	4.160	4.039	3.605	3.416	3.327	3.242
8	5.335	4.487	4.344	3.837	3.619	**3.518**	3.421
9	5.759	4.772	4.607	4.031	3.786	3.673	3.565
10	6.145	**5.019**	**4.833**	4.192	3.923	3.799	3.682

With the use of a table of discounting factors (Table 7.13) and a graphical representation (Figure 7.17), this leads to an IRR of 23%.

2. Furnace B provides varying cash flows:
   □ Investment sum: $1,600,000
   □ Utilization period: 10 years
   □ Annual revenue:
   □ Year 1–6: $1,150,000
   □ Year 7–10: $1,380,000
   □ Annual costs:
   □ Year 1–6: $550,000
   □ Year 7–10: $660,000
   □ Overhead costs: $300,000

For this case, the calculation has to be done by iteration, starting with a realistic value of the interest rate. If the discounted cash flow is lower than the investment sum, the calculation has to be repeated with a lower interest rate. Let us start with a 16% interest rate in the first iteration. The annual net income I is for

$$\text{Years } 1-6 : I = 1,150,000 - 550,000 - 300,000 = \$300,000$$
$$\text{Years } 7-10 : I = 1,380,000 - 660,000 - 300,000 = \$420,000$$

Table 7.13 shows the discounting factors for the interest rates and number of years relevant for this example. Using this table and assuming an interest rate $k = 16\%$ gives Table 7.14.

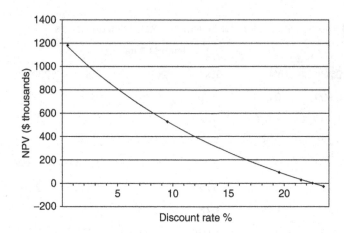

FIGURE 7.17 NPV as a function of discount rate.

Table 7.14 Discounting Factors (DFs) at $k = 16\%$

	DF$_1$	DF$_2$	DF$_1$ – DF$_2$ = DF	Cash	DF × Cash
Year 1–6	3.685	0	3.685	300	1,106
Year 7–10	4.833	3.685	1.148	420	482
Discounted capital at $k = 16\%$					1,588
Minus investment sum					–1,600
Equals missing value					–12

An analogous calculation with $k = 15\%$ yields a surplus of \$54,000. Comparing the two results shows that the actual interest rate is closer to 16%. We can now compare the two furnaces and see that *furnace A with an IRR of 23% is clearly more attractive than furnace B with an IRR of 16%.*

Profitability of Investment Projects

Investment appraisal methods can be considered a quantitative element in evaluating an investment and help make the consequences of uncertainties in the future more transparent. Table 7.15 shows as an example numbers for some typical investment projects.

Risk and Uncertainty

Nearly all management decisions are made in an atmosphere of *uncertainty*, because it is never possible to predict the future with complete confidence. Although they are

Table 7.15 Maximum Payback Periods and Minimum Internal Rates of Return		
Type of Project	**Max. Payback Time (years)**	**Min. IRR (%)**
Machines for mechanical manufacturing	3	30
Machines for electronic manufacturing	1.5	60
IT equipment	1.5	60
Product rationalization (power technology)	2	50
Product development (power technology)	3	20
Product development (electronics)	2	40

closely related, a distinction is made between *risk*, which can be assessed on past experience and insured against, and *uncertainty*, which is not insurable.

Estimates of a project's future cash flow or future costs are inevitably *uncertain*. Any single measure, such as NPV calculation, is an unreliable guide to its worth, as it does not contain a guide to the degree of uncertainty attached to it. Also there is always a risk that the actual return will differ markedly from the expected return.

Formally, consideration for such risks in the investment calculation can be made in several ways:

- Allow for a higher value of the expected profit by a shorter payback period or higher assumed revenue.
- Use higher values for the cost of investment through higher annual costs or higher costs of development (e.g., when 50% of the development projects have to be stopped, it is safer to double the development costs in the investment analysis).
- Lower the expected cash flow by decreasing the annual profits. For example, when there is a 50% probability of the reflow of cash going down to 0, work with only 50% of the reflux of capital.

Failures in investment decisions are frequent, though no one likes to talk about it. People mostly prefer to talk about successes.

Case Study. ZnO Varistors: NPV Comparison Between In-house R&D and Licensing

We discussed the case study on ZnO varistors in Chapter 3. Based on the qualitative information shown in Figure 7.18 and a combination of assumptions and guesswork based on some real data, we can use NPV calculations to see which approach pays off better.

Assumptions were made about the cost of the pilot plant, the production plant (not shown in detail), the labor force, and administrative costs. Using an internal

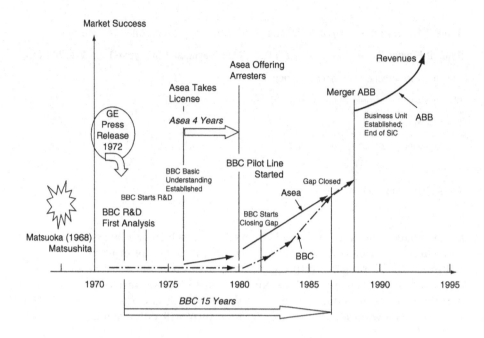

FIGURE 7.18 Historical comparison of varistor business development at Brown Boveri and Asea.

interest rate for the investments, the yearly costs could be calculated. The same assumptions were applied to the amount of money, which Asea had paid as a license to Matsushita. Using a graphical method the yearly incomes were estimated as a function of time. Without going into these details, six variants were calculated.

NPV (in million CHF)

1. Brown Boveri (1976–1998) with consideration of costs of R&D at corporate research: 0.4.

2. Brown Boveri (1976–1998) without costs of corporate research: 2.8.

3. Brown Boveri (1980–1998), without costs of corporate research. 1980 was start of pilot plant: 4.9.

4. Brown Boveri (1980–2000) without corporate research: 5.9.

5. Asea (1978–1998) with license: 2.3.

6. Asea (1978–2000) with license: 3.2.

Discussion

If corporate research is 100% funded by headquarters, then the varistor business unit faces only a "tax," regardless whether it uses corporate research or not—the service is "free." From the viewpoint of the business unit, this leads to a better value for NPV variants 1 and 2. If the business unit has to contribute to the funding of the project at corporate research (mixed funding), then the NPV would be of less value.

If we disregard corporate research (because the work is "free"), then the project in the business unit starts only in 1980 (date of pilot plant investment by business unit).

The NPV then increases further (variants 3 and 4).

Variants 2 and 5, as well as 4 and 6, show a comparison between Asea and Brown Boveri, always looking at the same final year. In both cases, Brown Boveri looks better.

Variants 2 and 6 are taken for utilization periods of 22 years. The results are similar, with Asea being slightly better. This is because they were moving faster.

Conclusion

Depending on the view we take, the results are either in favor of Brown Boveri or Asea. The learning for Brown Boveri is that the combination of "free" corporate funding and in-house research is basically a better approach than licensing by Asea.

On the other hand, Asea was always faster and so looks good over a longer time period. The fact that Brown Boveri waited until 1981, then took only about four to five years to close the gap with Asea is also interesting—money invested later is worth less, although the same has to be said about the delayed stream of revenues. Had Brown Boveri been faster in setting up the pilot plant and production, then it would even look better—*speed is most important*.

Analogy to Government Funding of High-risk Projects

Let us assume a company is planning the transfer of R&D results from public research institutions into its own organization. Further public funding for work at a public research institution makes this transfer very cheap. The company, as opposed to a business unit in Brown Boveri, need not pay a "tax" to fund the public research institute. If the researchers are successful and show good initiative and leadership, then the NPV of such a development could be even higher than the ones computed earlier.

SUMMARY

In financial accounting, we take a top-level view of where the company is at a particular moment of time: the balance sheet. While this is mostly important for people outside the company, such as investors, and top-level management, who have to assess the cash position and make decisions concerning the future development of the corporation, other measurements also are important for internal management. The income statement is a retrospective view about the performance of the company—where the revenues came from and the reason for the costs—leading to the view into the future: the cash-flow budget. Most important is a continuous view on the amount of cash left in the company—will it be enough to finance growth or is there a possibility that it may run out of cash and go bankrupt? A very good source of information is the annual report of a company, which provides information about a company from all possible angles of view: technology, business partners, markets, cash flow, risks, major opportunities, management team.

Managerial accounting deals with methods to calculate costs, revenues and profit—and what measures have to be taken to raise revenue, lower costs, and increase the profit. While calculating fixed and variable costs for just one product is simple and straightforward, more uncertainty and degrees of freedom are introduced when dealing with many products. The most important learning is marginal costing, where the difference between revenue and variable cost is the contribution to the fixed costs. The art is to find ways to make the sum of all contributions larger than the fixed costs, because only then are profits generated.

Several methods of investment calculations are discussed. In the static methods, no time value of the money is considered. More sophisticated methods consider the time value of money as well, and the best-known investment calculation method is the net present value calculation. In a case study of varistors, this method is used to compare two methods to develop a business: in-house R&D and licensing. The model shows that, by small variations in the input parameters, almost any result can be obtained, which should serve as a warning signal not to give the NPV method too much importance in assessing the importance of R&D projects.

REFERENCES

[1] R. Follett, How to Keep Score in Business, Follet Publishing Co., Chicago, 1987.

[2] P.C.F. Crowson, B.A. Richards, Economics for Managers, Edward Arnold (Publishers) Ltd., London, 1975.

[3] C. Barrow, Financial Management for the Small Business, Kogan Page Ltd., London, 1985.

The Product/Market Matrix

8

OBJECTIVES

The objective of this chapter is to make you familiar with the growth matrix, showing various ways how a business can evolve:

- **Market penetration**. This is a very common form of business growth in established businesses: products, technology and market are known.
- **Product development**. This is also a common growth pattern for materials-based businesses. New technologies permit a change in the product design or newly developed products for an established market. Several case studies address this variant: high-temperature superconductors, La-Mo cathodes for radio transmitter valves, and the use of nanotechnology in textile fabrication.
- **Market development**. In this growth model, the products are established and growth occurs by finding new markets. The case study American Super-conductors, although discussed under "product development" in comparison with ABB, has some elements of new market development.
- **Diversification**. This is definitely the most challenging way to grow a business, as will be shown in the example of the evolution of the LCD business.

GROWTH MATRIX

If we define technology in materials as the accumulated knowledge in a class of materials, consisting of the know-how relating to a certain range of compositions, microstructures, and physical and mechanical properties, as well as the know-how in processing these materials, then we can visualize the next step—developing a product.

Examples of new technologies are precision-cast single-crystal superalloys, carbon nanotubes, zirconia ceramics, and carbon-fiber-reinforced polymers. Basic and applied research are the tools that generate the knowledge to bring these technologies to a standard from which existing products can be improved and

181

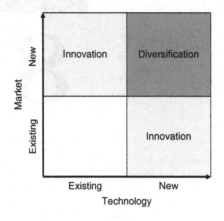

FIGURE 8.1 Technology/market matrix.

FIGURE 8.2 Ansoff's growth vector: The product/market matrix.

completely new product developed. In the beginning of the innovation process, very often a multitude of potential product ideas drive the process. A new product idea carries with it a certain likelihood of success or failure. This uncertainty applies to both existing companies and completely new startup companies built around a new product idea.

As part of the overall strategy formulation process, a company also needs to decide what directions to choose with regard to products and markets. There are several ways to approach the issue. As materials technologies often are the source of new product ideas, we can bring all conceivable variants of development into the form of the technology/market matrix (Figure 8.1).

As new technologies and new market needs are the main drivers for product development, a similar way to structure the evolution of new products, markets and businesses is the Ansoff growth matrix [1] (Figure 8.2). The Ansoff growth matrix is a tool that helps businesses make decisions on their product and market growth strategy.

The output from the Ansoff product/market matrix is a series of suggested growth strategies that determine the direction for a business strategy: market penetration, new product development, market development, and diversification. These are described in the remainder of the chapter.

MARKET PENETRATION

Market penetration is the usual playing field of existing companies. They can follow three strategies:

- Withdrawal from the business, once the managers recognize that it is no longer competitive or able to sustain a profitable business,
- Consolidation of the business by taking all the required steps to remain cost competitive,
- Market penetration, as a means to increase the market share of current products. Usually this is accomplished by process improvements, leading to lower product costs; other strategies can involve lowering prices or greater efforts in sales promotion and marketing. In a growth market, sales can be increased without increasing the market share. Penetration in a static or declining market is much more difficult and requires much more effort.

In general as the company is dealing with products and markets it already knows, the risks involved are lowest. The danger is that it sometimes promotes a mainly risk averse culture.

It is beyond the scope of this book to go into more practical details on the matter.

NEW PRODUCT DEVELOPMENT

Two possible strategies of new product development exist:

- Development of a new product with existing technologies and competencies.
- Development of new technologies and new competencies in existing markets.

Increasing the number of turbocharger variants following the market needs was a strategy based on existing technologies. Switching from a two-piece Al alloy impeller to a one-piece, complex-shaped one for use in industrial turbochargers required a new design and a focus on flexible manufacturing processes. Introducing thermal barrier coatings in gas turbine components was an important step forward in raising the turbine inlet temperature and a substitute for the more-expensive single-crystal alloy technology. It required new competencies to engineer and apply the coatings.

Most of the risks are connected to the introduction of the new technology. An example is the challenge the discovery of high-temperature superconductors brought to companies that had products based on electrical conductors, such as motors, transformers, or generators. The case study on high-temperature superconductors describes the different ways established and startup companies dealt with this challenge.

Case Study. High-Temperature Superconductors: Technology Push

Introduction

High-temperature superconductors (HTSCs), when discovered in 1986 [2], presented a huge challenge but also opportunity to companies like ABB, GE, and Siemens to radically improve their products. This case study traces the development of large-scale applications, such as superconducting fault current limiters (SCFCLs), at a large, established company (ABB) and compares it with the evolution of a startup company, American Superconductors (AMSC). ABB, having reached a high level of competence, basically had withdrawn from the project (cooling costs too high), while American Superconductors invested aggressively in the manufacturing of HTSC in new products and new fields, like power electronics. A comparison of the two approaches shows different ways of managing innovation (flexibility toward business goals and different ways of taking risks).

R&D at ABB

In 1986, HTSCs, cooled by liquid nitrogen, were discovered by Bednorz and Müller [2] at the IBM Research Center in Switzerland. From 1987 to 1993, the critical temperature, T_c, was further increased from 92 K ($YBa_2Cu_3O_x$/YBCO) to 130 K ($Hg_2Ba_2Ca_2Cu_3O_y$). Further developments led to more practical HTSCs: at first Bi-2223-, then YBCO-123-based coated conductors.

ABB, like many other companies, started an aggressive research program [3] and, together with a business unit, developed a series of real-life demonstration units (Figure 8.3).

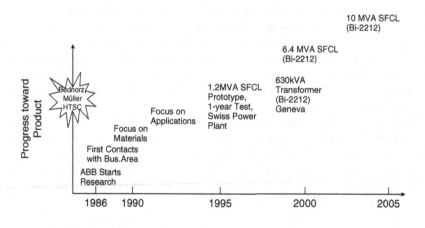

FIGURE 8.3 History of development of HTSC at ABB.

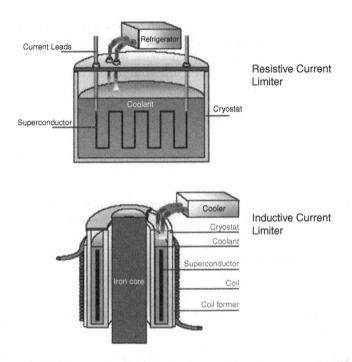

FIGURE 8.4 The two concepts of SFCL.

While the initial research focused on further material and process development, various potential applications came into consideration, leading to a focus on superconducting fault current limiters.

Figure 8.4 shows two possible concepts of fault current limiters. Basically, the idea is to transform the superconductor to a normal electrical conductor by heating the material through high currents above the transition temperature.

The simplest superconducting limiter concept, the resistive current limiter, exploits the nonlinear resistance of superconductors in a direct way. A superconductor is inserted in the circuit. During a fault, the fault current pushes the superconductor into a resistive state. The superconductor in its resistive state can also be used as a trigger coil, pushing the bulk of the fault current through a resistor or inductor. The advantage of this configuration is that it limits the energy that must be absorbed by the superconductor.

Another concept uses a resistive limiter on a transformer secondary, with the primary in series in the circuit. A copper winding is inserted in the circuit and coupled to an HTSC winding. During normal operation, a zero impedance is reflected to the primary. Resistance developed in the HTSC winding during a fault is reflected to the primary and limits the fault.

The inductive limiter can be modeled as a transformer. The impedance of this limiter in the steady state is nearly zero, since the zero impedance of the secondary (HTSC) winding is reflected to the primary. In the event of a fault, the large current in the circuit induces a large current in the secondary and the winding loses superconductivity. The resistance in the secondary is reflected into the circuit and limits the fault.

Critical Issues
Several hurdles have to be overcome to make this development a success.

Conductor Cost
If we take the cost of Cu wires as a reference, then high-temperature superconductors are much more expensive, but a factor of 5 seems to be attainable for YBCO-coated conductors:

Cu	10–25 USD/kAm
NbTi3	5–6
Bi-2223	200 → 50(?)

For YBCO-coated conductors, with their low cost of raw materials, a target factor of 5 is feasible.

More Serious Hurdle   The main cost hurdle was the cooling system. It was not a new technology—it had been around for a long time—but was still very expensive. Were there new, lower-cost technologies for cooling? The next question concerned the performance/cost ratio of the superconductor. Bi-2223 had been the most established superconductor, but the high price of the needed Ag sheath material undermined the cost. YBCO-coated conductors were the most obvious candidates, but a low-cost production technique still had to be established.

Market Input
Basically, a SFCL is a low-cost fuse. To be successful in the high-voltage (HV) market, a cost target of 50–100% of HV switchgear had to be met. Its value corresponded to that of the transformer that would be saved plus the load switch. The market potential was certainly interesting:

HV switchgear: World market 100,000/year
HV transformer for network coupling: 100/year

The cost of cooling had an effect on which technology to choose. At high power levels, it had become possible to make the resistive SFCL cheaper than the inductive one (Figure 8.5).

As the figure shows, the inductive current limiter would be more attractive for low power, because no current leads are needed, whereas the resistive current limiter would be more attractive for higher power, because thin conductors can be realized only with resistive current limiters.

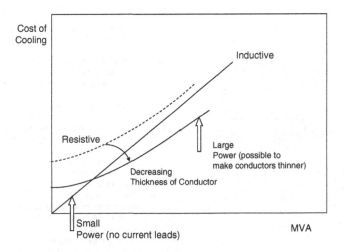

FIGURE 8.5 Relatively lower cooling costs for resistive SFCL at high power.

Cost Analysis and Potential Applications

When the search for potential applications started, a very systematic approach was taken. Noe and Oswald [4] systematically looked at the cost/benefit scenarios for each application, using complex net present value calculations. They came to the conclusion that the two most likely applications were generator connection and the coupling of local generating units, like wind power stations.

Looking at the whole system of an electric HV network, 11 potential applications were identified (Figure 8.6).

Economical Evaluation Method [4]

An economical evaluation method starts with the assumption that no costs of the SFCL are known and concentrates on the prospective savings by using a SFCL. The method is based on the present worth method or net present value NPV (for more details on NPV, see Chapter 7).

The present worth of savings PW_s originates with costs for conventional methods for short-circuit current limitation, lower losses, delay of improvement, or lower dimensioning of installations and devices. It is set equal to the NPV of the costs by applying a SFCL, PW_{SFCL}:

$$PW_s = PW_{SFCL}$$

The NPV of SFCL costs is given by the sum of NPV of purchase price PW_P, the NPV of the cost of losses PW_L, and the NPV of the maintenance costs PW_M:

$$PW_{SFCL} = PW_P + PW_L + PW_M$$

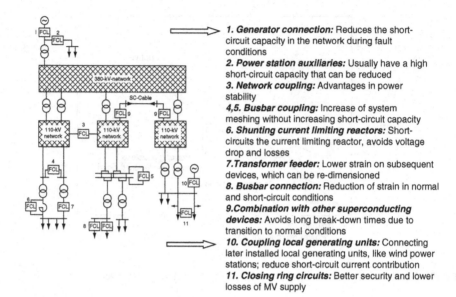

1. **Generator connection:** Reduces the short-circuit capacity in the network during fault conditions
2. **Power station auxiliaries:** Usually have a high short-circuit capacity that can be reduced
3. **Network coupling:** Advantages in power stability
4,5. **Busbar coupling:** Increase of system meshing without increasing short-circuit capacity
6. **Shunting current limiting reactors:** Short-circuits the current limiting reactor, avoids voltage drop and losses
7. **Transformer feeder:** Lower strain on subsequent devices, which can be re-dimensioned
8. **Busbar connection:** Reduction of strain in normal and short-circuit conditions
9. **Combination with other superconducting devices:** Avoids long break-down times due to transition to normal conditions
10. **Coupling local generating units:** Connecting later installed local generating units, like wind power stations; reduce short-circuit current contribution
11. **Closing ring circuits:** Better security and lower losses of MV supply

FIGURE 8.6 SFCL in power systems (© 1999 IEEE).

The mathematics used in the calculation is fairly complex and will not be further discussed here. Figure 8.7 shows the results for the various applications, indicating that higher prices are possible for the two applications mentioned previously.

Another consideration for potential sites of such applications is the question whether the component replaces existing, more-expensive solutions or will be installed in a completely new network. Thus, advantages can be seen when building a completely new HV network (e.g., in China): We can forget about all measures to control short circuits, which we need in an existing network:

- HV switchgear.
- Mechanical design of network components (e.g. transformers and conductors).
- Thermal design of components.

Potential Ways to Increase Performance and Lower Costs

1. Looking for applications that have a higher valuation of performance (in new networks, e.g., in China, or new applications)
2. Higher-performance applications (i.e., better current limitation, e.g., max current/nominal current, or recovery time).
3. Lower costs with
   □ Higher critical current density → less superconductor material.
   □ Thinner conductors → less ac losses.

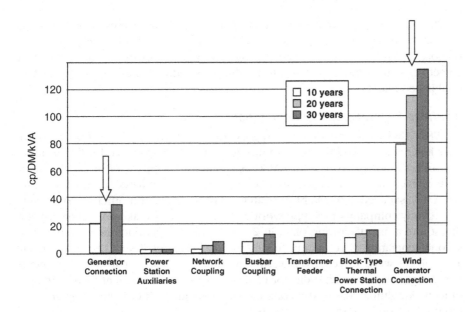

FIGURE 8.7 Medium specific purchase price of SFCL for life-time assumptions of 10, 20, 30 years [4] (© 1999 IEEE).

> □ Better contact to electrical bypass → fewer hot-spot problems → shorter conductors, higher mechanical strength, cheaper cooling.
> □ A process to manufacture a multilayer superconductor.

Conclusions

Since its invention in 1986, high temperature superconductors underwent important improvements. The critical temperature, T_c, was increased from 92 K ($YBa_2Cu_3O_x$/YBCO) to 130 K ($Hg_2Ba_2Ca_2Cu_3O_y$). Further developments led to more practical HTSCs: at first Bi-2223-, then YBCO-123-based coated conductors.

It is still difficult to forecast whether SFCL can be produced at low enough costs. However, in the medium term, it seems possible that some locations are economically attractive enough to use SFCL.

Feeder locations of power stations and wind generators are the most economical places for SFCL. The main requirements for an economical usage of SFCL are *low losses and a long lifetime*.

Cooling costs are still considered the main obstacle.

The example given for ABB is representative for all major companies engaged in electric power businesses. Major progress has been achieved mostly where strong government funding was available, but no products have been introduced into the commercial market.

The following example shows that small startup companies, which have all their attention focused on just a few major goals, follow a different path, which is more difficult to predict than the one in large companies, where the main objective is substitution of existing products by radically new ones, but serving established markets.

American Superconductors Corporation

In 1987, AMSC was founded by Gregory Yurek, a former MIT professor. Shortly after its creation, I had the opportunity to visit the company, and ABB Corporate Research started to work with AMSC as a potential supplier of material. Due to slow delivery and high costs, the cooperation was terminated after a few years, but AMSC succeeded to establish strategic partnerships with other global players. As the only company of its type worldwide, it decided to take on a huge risk and invest into a manufacturing plant in 2003 to produce conductors, switching into second-generation wire manufacturing in 2007 (Figures 8.8 and 8.9). Practically all projects were driven by government funding (U.S. Department of Energy, U.S. Navy), very often an indicator of slow progress and frequent failure. In spite of its major annual losses, the company had enough cash to keep going and continue taking on high-risk investments.

Search for New Markets and Capabilities

AMSC kept its eyes open for opportunities to either find new markets or reduce its risk. The company, which had started out to produce HTSC wires, began to look for products using the wires and widened the potential market to all applications where high electric power is needed. This includes

- Power cables.
- Motors, generators, and synchronous condensers.
- Maglev trains.
- Transformers.

FIGURE 8.8 AMSC's plant to produce HTSC wires [5].

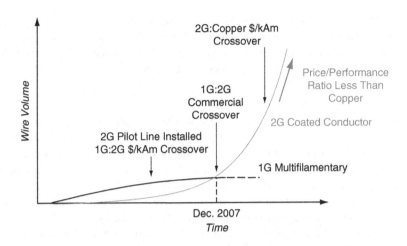

FIGURE 8.9 AMSC's plan to switch from first- to second-generation wires.

- Fault current limiters.
- Military applications.

Under the assumption that there would be a growing demand for HTSC wires, the price/performance ratio would become lower than that of Cu. This will have to happen to move the business from demonstration units to commercial products. The commercial uses of superconductor products are still limited, and a widespread commercial market may never develop.

AMSC recognized that it could acquire other, more mature technologies, and this would lower the risk remaining in the HTSC business while helping to develop the next generation market for HTSC applications.

AMSC diversified into two such technologies by acquiring two companies, one dealing with dynamic reactive power solutions and the second with wind power systems. Very quickly, this decision proved to be the right one, and the major business is now based on power systems solution, generating a profit and helping fund the further development of a second generation of HTSC wires.

Another impressive element in AMSC's business strategy is to establish a large portfolio of patents and licenses, about 700 in 2007, which is used successfully in licensing its technology to other companies with more market access.

Typically, for a material-based business, at least 20 years is required until net profit is reached. Once this hurdle has been taken, a business may grow very rapidly. In AMSC, about 90% of the revenue in 2007 had been generated by the two acquired companies, which produce conventional products but with a long-term potential to use HTSC products. Revenue, mainly from the wind power business, has been growing at a very fast rate. Although a positive gross margin has been reached, there continues to be a substantial net loss due to the continued high investment in the R&D of HTSC products. There, all revenues come from contracts, which are partially subsidized by government grants. Still, the company remains optimistic and focused on the long-term goal of entering the commercial market with HTSC products, which may be in applications not considered about 10 years ago, supersecure HV networks, for example.

Information from AMSC's Annual Report

AMSC's annual reports are excellent examples of the various changes in strategies and financial performance of the company. Figure 8.10 shows, retrospectively, how the initial source of revenues gradually changed from government-funded contracts to commercial ones, a strong indication of the growing strength of the company. Table 8.1 shows a multiyear overview of key financial data.

The reports show that, in spite of continuously high yearly losses, the company never was in danger of running out of cash, which can be seen as a sign of confidence on the side of the investors. The second conclusion is the growing importance of the business coming from the new AMSC power systems division, whereas the high loss on the side of AMSC superconductors is a proof of management's commitment to continuing investment in the manufacturing processes of HTSC wire and potential new products using these wires.

The annual reports also give very clear explanations of past business transactions. The 2008 report provides a good description of the acquisition of two companies in 2007. These companies helped generate fast-growing revenues. The AMSC superconductors division continues to focus on developing the wire manufacturing technology and various HTSC products: power cables, fault current limiters, and secure super grids.

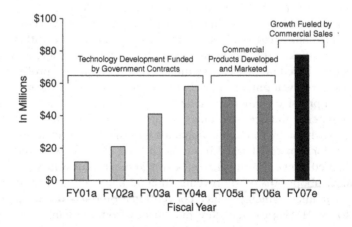

FIGURE 8.10 Change from government-funded business to commercial products business, AMSC's actual and estimated revenue.

Table 8.1 Financial Data from Annual Reports of AMSC (for Years Ending March 31; in Thousands, Except Per-share Data)

	2008	2007	2006	2005	2004
Revenue	$112,396	$52,183	$50,872	$58,283	$41,309
Net loss	(25,447)	(34,675)	(30,876)	(19,660)	(26,733)
Net loss per share	(0.65)	(1.04)	(0.94)	(0.70)	(1.10)
Total assets	261,234	132,433	133,470	158,917	129,899
Working capital	124,334	34,942	66,220	77,272	46,202
Cash, cash equivalents, and short- and long-term marketable securities	106,323	35,324	65,669	87,581	52,647
Stockholders' equity	208,452	101,621	115,100	143,510	115,452
Revenue					
AMSC Power Systems		$30,850	$15,001		
AMSE Superconductors		$21,333	$35,871		
Total		$52,183	$50,872		
Operating Income (Loss)					
AMSC Power Systems		$402	($3,641)		
AMSE Superconductors		($3,419)	($27,549)		
Unallocated corporate expense		($5,515)	($2,297)		
Total		($36,532)	($33,487)		

Learning Points

AMSC is a good example of successfully setting up a new materials-based business. High-temperature superconductors are clearly a very high-risk group of new materials, and it requires very high multiskill levels to successfully develop a potentially cost-effective method of wire fabrication. Maintaining focus over more than 20 years on the same idea has been a very important motivation in view of continuing substantial yearly losses.

These types of materials, in spite of their huge long-term potential, also contain great risks, and government funding has been vital to maintain the momentum of development.

To be successful in acquiring government support requires the ability to network across government institutions and find the right partnering companies with capabilities required to come up with a product.

It is of utmost importance to invest all the money needed to build up a substantial base of intellectual property rights. The success of AMSC and the global interest in this technology has led to a host of potential followers and competitors. To survive and even benefit from a licensing business requires leadership in intellectual property.

Several strategic partnerships with major players, such as GE, Siemens, Mitsubishi, and many other companies, help AMSC to share the risk.

Good communication with the outside world helped secure an increasing amount of cash flow into the company, which is necessary in the lengthy startup phase required for this type of business.

Comparing the development of HTSC at ABB and AMSC, we must ask the question, Will HTSC products be among the 5% of radical and successful innovations or among the 95% of failed innovations? The case study is a very good example of the important role startups play in pioneering high-risk product innovations.

The next case study is another example of the challenges connected with the development of a new technology to be used in an existing product.

Case Study. Lanthanum Oxide–Molybdenum Cathodes: Incomplete Target Specification

Radio transmitter valves were a very profitable business of Brown Boveri in the 1970s. The high-power vacuum tubes consisted of a cathode made of thoriated tungsten, an anode, and a grid. Under high voltage, the cathode, when heated to high temperatures, emitted electrons, providing the current required for the operation of the transmitter valve. It became known through the research of Lafferty [6] that cathodes containing lanthanum had a great potential to replace thoriated tungsten, because the energy required for electron emission was significantly lower. So the BBC Corporate Research Center was approached by the business unit with the request to develop such a cathode with a defined emission current and a lifetime of 10,000 hours.

From Specification to Concept

The starting specification was to demonstrate the feasibility of a material combination still to be found. The material to be finally used was La_2O_3 dispersed in molybdenum. The surface of the molybdenum was carburized into Mo_2C. The function of the molybdenum carbide was to reduce the lanthanum oxide during operation of the cathode into lanthanum, which then diffused to the surface. The rate of evaporation of La at the surface was quite high, which limited the lifetime of the cathode. From the theory of activated sintering of tungsten [7], it was known that small additions of Ni increased the grain boundary diffusion of W by several orders of

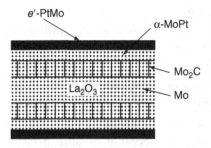

FIGURE 8.11 Schematic view of a La-Mo-Pt-cathode.

magnitude. It was now speculated that using another element from the same group of the periodic system as Ni might have a similar effect of enhancing grain boundary diffusion. The element chosen was Pt. The hypothesis was that Pt would not only accelerate the grain boundary diffusion of molybdenum, but also the diffusion of La, and the hypothesis was proven correct. Figure 8.11 is a schematic cross-section through a La-Mo-Pt cathode after about 100 hours of service [8].

Figure 8.12 shows the relative improvement of the emission current density of a lanthanum-containing cathode compared with the established thorium-containing tungsten cathode [9].

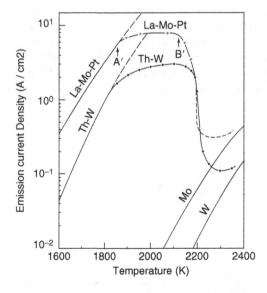

FIGURE 8.12 Emission current density of La-Mo-Pt sintered cathodes.

Manufacturing Challenge

Tests were done quickly on as-sintered samples, proving that the original hypothesis was correct. But it was still a long way to a product. Two major tasks still needed to be completed:

- Finding the right process to manufacture wires from a sintered perform.
- Obtaining a lifetime of 10,000 hours at the operating temperature.

Both tasks were extremely challenging. A visit to one of the leading manufacturers of tungsten wires was unsuccessful, because the as-sintered preform was still too brittle to be deformed by hot swaging. Hot swaging is the initial manufacturing step in producing thin-diameter rods, which can then be further reduced in diameter by wire drawing. Through other contacts, a drop forging hammer was found at DFVLR, a German research center, which allowed extruding the preform at temperatures around 1500°C. This produced a bar with enough strength to withstand the deformation during the subsequent swaging process. Now, the same processes of platinizing and carburizing were applied as in the initial sintered samples. The target to reach a lifetime of 10,000 hours, though, could not be reached.

Lifetime Optimization

To find ways to increase the lifetime, a model based on the diffusion of both lanthanum and platinum to the surface of the cathode had to be developed. If there was enough lanthanum oxide to be used as a supply of lanthanum, part of which evaporated from the surface of the cathode, then the lifetime limiting step was the availability of enough platinum. The theoretical physics group of the BBC Research Center developed a two-layer diffusion model leading to the optimum thickness of a Pt-rich core from which Pt could diffuse radially outward, and maintain a life of 10,000 hours [10]. To get the maximum amount of Pt in the core, a thin Pt wire was placed at the core of the extruded Mo–lanthanum oxide bar. The composite material was then hot swaged, resulting in a wire containing Pt in the center (Figure 8.13), and the final experiment led to the desired lifetime.

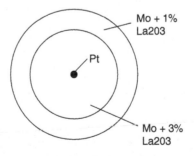

FIGURE 8.13 A two-layer model consisting of various compositions of molybdenum and lanthanum oxide activator plus an inner core of diffusion-enhancing Pt.

Other variants consisted of adding Pt powder to the inner core or providing a Pt layer dividing the two layers.

When the cathode was introduced into the vacuum tube for a final product test, it turned out that the lanthanum evaporating from the surface of the cathode condensed on the grid wire, negatively affecting the switching behavior of the vacuum tube. The motivation of Brown Boveri to continue the research had also disappeared, because the entire business was sold, since it no longer fit the core businesses of the company.

Lessons Learned

- The research results were outstanding, and the entire project almost became a success.
- There was, from the beginning, very close cooperation between the scientists in the research center and members the business unit, and the skills were complementary.
- It was known that condensation of La on the grid was a factor to be considered, but before the start of the research program, it was assumed that the problem was not significant. There was no hint in the target specification to consider this a problem.
- Before completion of the entire project, the company disengaged itself from that business by selling the business to an outside company. Because of the newly surfaced problem with the grid, the development work was not continued. This can happen during any development and is unpredictable.

The following case study describes a success story of a medium-sized company that introduced a challenging new technology without the help of an internal R&D department.

Case Study. Nanosphere®: Nanotechnology in Textiles by Subcontracting R&D (Personal Communication with Norbert W. Winterberg, COO of Schoeller AG)

Big innovations often come first from small- and mid-sized companies. A good example is Schoeller Textile Group, a subsidiary of Albers & Co. In 1849, Rudolph Schoeller started textile manufacturing in Silesia, Germany. In 1867, his sons Arthur and Caesar established the company Schoeller & Sons in Zurich, which in 1979 became Albers & Co. At that time, many textile companies in Switzerland faced big problems and competition from low-cost producing countries such as India and China. The company realized that the only way out of this dilemma was to become more active in creating new technologies at a high speed. In 1992, a high-performance textile factory was constructed in Sevelen, Switzerland.

The company established joint venture companies in Hong Kong and Bombay, which used newly developed technology for low-cost production, but the Swiss company always managed to maintain a lead on critical parts of the technology.

In addition, sales companies in all parts of the world provided access to the global market.

Schoeller Technologies AG

An interesting concept emerged. To inject more dynamics into development and bring it closer to the business, Schoeller Technologies AG was established as a daughter company of Schoeller Textil AG in 2004.

The main business idea was to *license technology*—to the parent company, a joint venture partner, or later nonpartner companies. Its business strategy contained four core elements:

- Conducting R&D projects, which are financed by Schoeller Textil AG. The company coordinates diverse research and development projects for futuristic textile innovations, taking into account a high level of added value and consumer-friendly handling.
- Managing and monitoring patenting, creating intellectual property.
- Licensing intellectual property and technology.
- Marketing: Sales, branding and marketing of Schoeller technologies worldwide.

The partnering model increased pressure to continuously innovate, and the company demonstrated that new developments could be done in a few months.

For its own research projects, the company relies heavily on externally accessible know-how, an approach contrary to what large companies usually do. Such know-how is accessible at universities or large government-funded research centers. Most of the time the task of Schoeller Technologies AG is to assemble the right international R&D team and guide the respective research projects by persons who are also market oriented. There is an ongoing interaction among the various functions of the company, and progress of each research project is closely monitored by textile engineers of the holding company, who make sure that not only the scientific targets are reached but also the manufacturing processes and market-relevant data are generated. Throughout the entire duration of a project, flexible practical tests can be carried out in addition in Schoeller Textil AG's state-of-the-art production plant. The case shows that parts of some functions can be accessed even when they are outside the organization of a company.

One of the most interesting examples of a new product development is Nano-Sphere®, which uses the so-called lotus effect.

The Lotus Effect as a Source of Innovation

The lotus flower (*Nelumbo nucifera*) is the symbol of purity in several Asian countries. The major reason for this is the self-cleansing property of its leaves. Even after emerging from mud, the leaves do not retain dirt when they unfold. This property has been studied intensively by the two botanists, Barthlott and Neinhuis from the University of Bonn. In 1975, they discovered the reason for this self-cleansing effect [11]. Before that, it was taken as intuitively obvious that the smoother is a surface, the less dirt and water adhere to it. By using a scanning electron microscope (SEM) the two scientists discovered that the surface of some lotus leaves showed a combination of nano- and microstructures that gave the surface a rough structure.

The explanation for the effect lies in two physical characteristics:

- One of the properties of these microstructures is to repel water.
- The nanostructures found on top of the microstructures are made of waxy materials that are poorly wettable.

The combination of the chemistry, the ultrastructures, and the adherence properties of dirt and water to the surface is what Barthlott and Neinhuis named the *lotus effect*. The physical basis of the lotus effect is the wetting angle of a droplet of liquid sitting on the surface of a material with a nanostructured rough surface. If the wetting angle is less than 90°, we consider the fluid a good wetting agent (hydrophilic). It would actually infiltrate the rough surface. The actually observed wetting angles are closer to 150–180°, which makes the liquid nonwetting (hydrophobic). The natural "lotus effect" is based on a three-dimensional surface structure wherein the wax crystals formed on leafs by self-organization account for a microroughness strongly promoting the self-cleaning effect of the plant.

A droplet on an inclined super hydrophobic surface does not stick to the surface. It also does not slide off; it rolls off. When the droplet rolls over small contaminating particles, the particle is removed from the surface if the force of absorption of the particle is higher than the static friction force between the particle and the surface. Usually, the force needed to remove a particle is very low due to the minimized contact area between the particle and the surface. As a result, the droplet cleans the leaf by rolling off the surface (Figure 8.14).

History of NanoSphere® [12]

Research started in 1998, with the goal to create hydrophobic and oleophobic surfaces; partners in this cooperation were Nanotech in Brooklyn and a mid-sized

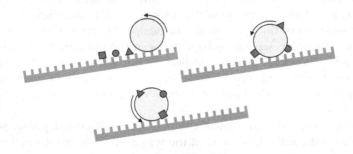

FIGURE 8.14 Self-cleaning effect of hydrophobic surfaces.

German chemical company with know-how in fluorochemistry. The first patent was filed in April 2000. The goal of oleophobic surfaces was not reached, however.

The second generation of research was done in cooperation with Texatech (in Wattwil near Zurich), which had close contacts with two professors at the ETHZ (both have since retired and their labs have been closed). This phase of research proved to be successful.

Further research was conducted in cooperation with Bühler and Partech. The goal was to improve permanence (avoiding abrasion). Also this goal was reached successfully.

An essential feature of NanoSphere is the use of a dispersion or emulsion system as a "guest/host" system, which enables a spatial self-organization of the finishing components. By this self-organization of the "guest" and the "host" components, an anisotropic distribution of the "guest" component or the dispersed phase within the "host" component is achieved within the finishing layer. In the final finishing layer, the guest component concentrates at the upper surface of the finishing layer and thereby dominates the physical, chemical, and physico-chemical properties at this phase boundary layer between the finishing layer applied and the surrounding atmosphere.

If gelling additives, such as high molecular-weight soluble polysaccharides or polar cross-linking components, like glycerol and methoxy methylolated urea derivatives, are added to the water phase of the dispersion system, a membrane forms on the tissue in addition to the aforementioned self-organization. As a result, the initially homogenous dispersion system partitions during drying into two liquid phases, referred to as *coacervate*s. One of these predominantly contains the gelling polymer fractions, while the other is dominated by the apolar, water- or oil-repellent components. Due to the cross-linking reaction during the drying process, the polymer gel contracts, leading to the formation of the pore system of a membrane out of the originally gellike structure.

The final finishing layer essentially corresponds to dispersion in the gel state. The heterodisperse system is utilized for the formation of columnar structures and thereby for the generation on the finished textile of a microrough surface exerting the lotus effect.

Self-organization and formation of membrane structures (i.e., the tendency to undergo partial phase separation of the guest and the host components) results in an accumulation of the hydrophobic or oleophobic guest components at the surface (i.e., the phase separation layer between the finishing layer and the surrounding air). Therefore, self-organization of the guest and host components results in dramatically enhanced water-repellent or oil-repellent finishing effects at the upper surface of the finishing layer as compared to a homogenously dispersed system.

Due to its properties, the guest component, or dispersed phase, is mainly responsible for the self-organization of the water-repellent or oil-repellent finishing layer (phase separation) and the formation of columnar structures having a directional orientation at the phase boundary layer and may consist of widely different but always very apolar water- or oil-repellent auxiliary agents, depending on the property profile of the finishing.

Specifically, the agents may be silicone oils, lipid modified esters, ethers, or amides (such as glycerol ester and ether, sorbitan ester and ether), being high-boiling-point, apolar liquids that diffuse toward the phase boundary layer (solid/gas) during the setting process and are set in a position promoting the water-repellent or oil-repellent finishing effect.

Another group of agents includes fatty esters, alkyl ethers ($C_{12}C_{25}$), and for example, polycondensed fatty amides, which are dispersed into the water-repellent or oil-repellent finishing emulsion in the form of solids and melt completely or only partially during the subsequent thermal setting and dominate the interface with their physical properties in accordance with the desired effect.

A third group comprises substances that form columnar structures. This group includes, for example, micronized waxes (particle sizes of 0.1–50 μm, preferably around 20 μm) such as polyolefin and fatty amide waxes as well as waxes that are lipid-modified aminoalkylation products, and hydrophobic silica particles (particle sizes of 5–100 nm), preferably nanoparticles having particle sizes of 5–50 nm, which are also dispersed into the water-repellent or oil-repellent finishing liquor and set in the finishing layer. Examples of such substances are Ceridust waxes (Clariant) or Aerosils (Degussa), which are preferably used.

Learning Points

- There are obvious advantages of working with R&D groups (at universities or government-funded research centers). We can save time and money, provided the company has the necessary communication skills. The consequence of working in teams with mainly outside groups is that it permits only personal contact; the know-how mainly remains with that group. Hence, there is a need to bring this often tacit knowledge into the company. This could be achieved by either buying a specialized company, building up know-how in-house, or making tacit information explicit. The critical know-how needed is textile technology and chemistry.
- With a new field, such as nanotechnology, we have to accept many failures and maintain focus and systematics.
- Because the project received government funding, the process was fairly bureaucratic, causing a slowdown. Speed is most important.
- Creating an R&D company within the holding company dramatically increased the flexibility and speed with which new product development was made possible.
- The licensing business became an incentive for more and faster development.

MARKET DEVELOPMENT

Market development refers to a growth strategy where the company attempts to sell existing products in new markets. These can be

- New geographical markets.
- New distribution channels.
- New pricing strategies to enter or create new market segments.

Well-known examples are new markets for existing (but also new) products in China and India.

The case study on American Superconductors, discussed together with R&D on HTSC at ABB earlier in this chapter, serves as an example, how, in the later stage of the evolution of this company, the acquisition of a wind power systems company opened the access to the wind power market for AMSC. Though still based on pre-HTSC technology, this new market access allowed AMSC to dramatically raise the revenues and make this new business one of the important sources of capital needed to continue long-term R&D on HTSC.

DIVERSIFICATION

Diversification involves bringing a new product to a new market. Inherently higher risks are involved with such a growth strategy, because the technology used in new products may still be untested and the company may have little experience in the new market. The attitude to diversification can vary greatly. To some it is a survival issue, because sinking prices of the existing products make it mandatory to come up with entirely new products. Others may see in a new technology a completely new opportunity.

In the case study about NanoSphere®, we already touched on the first applications of nanotechnology in textiles. It can easily be expected that nanotechnology will lead to many potentially completely new applications and new markets in high-performance textiles. Some of the topics mentioned in the published literature [13] are

- Nanotechnology in manufacturing composite fibers:
  - Carbon nanofibers and nanoparticles.
  - Clay nanoparticles.
  - Metal oxide nanoparticles.
  - Carbon nanotubes.
  - Nanocellular foam structures.
- Nanotechnology in textile finishing: Upgrade of chemical finishes and resultant functions.
- Nanoparticles in finishing.
- Self-assembled nanolayers.

Future development of nanotechnologies in textiles will focus on two items:

- Upgrading existing functions and performances of textile materials.
- Developing smart and intelligent textiles with totally new functions.

The new functions with textiles to be developed include

- Wearable solar cell and energy storage.
- Sensors and information acquisition and transfer.
- Multiple and sophisticated protection and detection.
- Health care and wound healing.
- Self-cleaning and repairing functions.

We can easily see that these new functions mostly represent both new technologies and new market opportunities. Nanotechnology certainly holds an enormously promising future for textiles. Like information technology, it will bring about a large market impact on new materials. While this conclusion seems obvious for a majority of the research community, no one can predict with certainty how it will come about.

While nanotechnology in textiles undoubtedly will play an enormous role in diversification opportunities, liquid crystal display technology can be viewed from the past.

Case Study. Liquid Crystal Display at Brown Boveri: A Too-Expensive Commercial Product Followed by Successful Licensing [14]

In the late 1960s, Brown Boveri decided to enter the field of medical electronics, which was then considered a promising growth area. Brown Boveri teamed up with the pharmaceutical company Hoffmann-LaRoche, which had the know-how in medicine and pharmacology Brown Boveri lacked. Since the visions of the two companies regarding the new business did not match, they came to the conclusion that further cooperation was not possible. Both companies, however, agreed that medical electronics would need much better display technology. Among various options, liquid crystal display (LCD) technology was considered the most promising. So research was quickly started at both companies. Brown Boveri had a further interest to use LCD technology in power plants to continuously show the operating status in a safe and clear format.

Liquid crystals were first discovered in 1888. They are substances that have the properties of both liquids and solid crystalline materials. For a long time, no one had an idea about applications. Only in 1963 did an optical effect became known that promised interesting technical applications [15]. A thin layer of the liquid crystal between two glass plates, under normal conditions, is clear and transparent. If electrodes are placed on the inside of the glass plates, the liquid crystal, under an applied voltage, creates a turbulence, which scatters the incoming light and makes the liquid crystalline layer nontransparent. This is the concept of a display module.

At the time, the success of the prevalent "dynamic scattering" (DS) technology was viewed with worldwide skepticism. The DS technology required higher operating voltages than the generally used low-power CMOS electronics and had severe optical deficiencies (a contrast strongly dependent on the surrounding light intensity). RCA, the early leader in the science of LCD, therefore started to cut back on research in the DS technology. This gave Hoffmann-LaRoche a chance to hire Wolfgang Helfrich [16]. The new "twisted nematic" effect could operate with much less power consumption than the DS technology. Optical contrast is independent of the type of surrounding light and operates under an acceptable range of angles. Further material development led to the discovery of liquid crystals, which could operate within an acceptable temperature range. After solving all the technical problems, the project was ready for transfer into production.

The first milestone was the exhibition of the world's first LCD-watch in 1973. All components of this watch were made by various Swiss companies. In 1974, after a short pilot phase, industrial manufacturing was started in a new plant in Lenzburg by Georges Keller, a very entrepreneurial person, who hired more top people. The first big customers were Japanese companies, especially Casio. But soon much of the know-how went to other Japanese companies, like Hitachi, that soon started to operate their own plants. The customers in the Far East had one big advantage over Brown Boveri: the first big markets were not watches but displays in pocket calculators, a market not accessible to Brown Boveri. The initial manufacturing technique required lots of manual labor, making the LCD displays from Switzerland expensive. The world market price for a digital display for a watch dropped from 6.80 CHF in 1977 to 1.50 CHF in 1980. To cut the price further, a second plant was built in Lenzburg in 1978, based on a supply contract with a large Swiss watch manufacturer, which, however, was never exercised.

Due to deterioration of the business situation, Brown Boveri started a 50/50 joint venture, Videlec AG, with Philips, in 1980. Philips, a semiconductor manufacturer, had all the required know-how to design and build chips for the control of large LCD elements. In addition, Philips had better access to markets for electronic components and intended to use LCDs in new product generations of HiFi equipment and telephones. Philips had started its own research and development of LCD elements, but could not catch up technically and in terms of manufacturing with Brown Boveri. Getting together with Brown Boveri in a joint venture would thus save it the investment in more research. The reason for Brown Boveri to enter into the joint venture was the delay by the Swiss watch industry to enter the digital watch market.

Videlec had to fight problems right from the beginning. Because of worldwide overcapacities, the price for a watch LCD-display dropped further to 0.50 CHF by 1981. Another problem came with the assessment of a new market: large-area LCDs with high resolution and low power consumption for laptops. This was made possible by the discovery of the "supertwist" effect in 1983 [17]. Both Hoffmann-LaRoche and Philips were initially skeptical about this potential new market. The CTO of Brown Boveri believed in it but had to realize that a larger investment in Videlec would be required, which Brown Boveri was not willing to make. He did make sure, however, that the patent rights would not be sold out prematurely. In 1981, the production in Lenzburg was stopped, also the one for large-area displays. The existing plant in Hong Kong continued to manufacture, but in 1984, Videlec was sold to Philips.

Both Brown Boveri's CTO and research director at that time, A. P. Speiser, assured that none of the top scientists had to leave the company. In addition, the business potential of licensing the patents for the supertwist effect and other LCD know-how was recognized. Over the years, license contracts were made with all major LCD producers, 90% of which were Japanese. As a result, a license income of several hundred million CHF served as a proof of the quality of the original research.

Learning Points

A lot was accomplished with a comparatively less effort than by Japanese competitors. Here are some of the guiding principles:

- Competent and motivated employees.
- A project embedded in a good scientific interdisciplinary environment.
- Open, mutual cooperation between corporate research and the business unit.
- Functional orientation of the LCD-research group.
- A thorough theoretical analysis showed early on the limits of the TN effect and led to the discovery of a successor technology, the "supertwist" effect.
- Although research and business were well integrated, the business did not become a success, because it did not fit the changed strategy of the corporation.

SUMMARY

A business can grow from a technology in four possible ways, described by the growth matrix:

- **Market penetration**. Here, efforts are undertaken to increase the market share for existing technologies, such as by increasing the efficiency of manufacturing or lowering the cost per product through higher volume production or manufacturing in low-cost countries.
- **New product development**. This is a frequently taken path by established companies to bring a new technology into an existing market. Several case studies are used to show some of the challenges: high-temperature superconductors at ABB, a large established power systems and automation systems company; La-Mo-Pt cathodes for radio transmitter valves at Brown Boveri; and the application of nanotechnology in textile applications at Schoeller AG, a medium-sized Swiss company.

- **Market development**. This relates to developing new markets for established technologies. The case study on American Superconductor Corporation, discussed together with ABB's work on the same topic, shows how growth of this startup company could be vastly accelerated by acquisition of two power systems companies, which potentially may use HTSC technology in the future, but facilitated market access to AMSC.
- **Diversification**. Combining the use of new technologies and new products in new markets definitely multiplies the risks in starting a new business. Brown Boveri, together with Hoffmann-LaRoche, were pioneers in developing LCD technology. To facilitate access to new markets, a joint venture with Philips was created, since an earlier attempt to work together with the Swiss watch industry failed. Switzerland, being a high-cost manufacturing site, turned out to be the wrong country in which to successfully establish this new business, which quickly moved to Japanese companies, which also were working on new applications based on LCD technology. Brown Boveri succeeded, however, in generating profits by licensing the technology worldwide.

REFERENCES

[1] I. Ansoff, Strategies for Diversification, Harv. Bus. Rev. 35 (5), (1957) 113-124.

[2] J.G. Bednorz, K.A. Müller, Possible high T_c superconductivity in the Ba−La−Cu−O system, Zeitschrift für Physik 64 (1) (1986) 189-193.

[3] W. Paul, M. Chen, M. Lakner, J. Rhyner, D. Braun, W. Lanz, Fault current limiter based on high temperature superconductors—different concepts, test results, simulations, applications, Physica C: Superconductivity 354 (1-4) (2001) 27-33.

[4] M. Noe, B.R. Oswald, Technical and Economical Benefits of Superconducting Fault Current Limiters in Power Systems, IEEE Transactions on Applied Superconductivity, ASC-9 (1999) 1347-1350.

[5] American Superconductor Corporation, available at www.amsc.com.

[6] J.M. Lafferty, Boride Cathodes, Physics Review 79 (1950) 1012.

[7] G.H. Gessinger, H.F. Fischmeister, A Modified Model for the Sintering of. Tungsten with Nickel Additions, Journal of Less-Common Metals 27 (1972) 129.

[8] G.H. Gessinger, C. Buxbaum, Der Einfluss von Platinmetallen auf das Emissionsverhalten von Lanthankathoden, High Temperatures-High Pressures 10 (1978) 325-328.

[9] G.H. Gessinger, C. Buxbaum, Evidence for Enhanced Grain Boundary Heterodiffusion, Materials Science Research 10 (1975) 295-303.

[10] C. Buxbaum, G. Gessinger, S. Strässler, German Patent DE 2822665.

[11] W. Bartlett, C. Neinhuis, The Purity of Sacred Lotus or Escape from Contamination in Biological Surfaces, Planta 202 (1997) 1-8.

[12] Available at www.nano-sphere.ch.

[13] L. Qian, J.P. Hinestroza, Application of nanotechnology for high performance textiles, Journal of Textile and Apparel, Technology and Management 4 (1) (2004) 1–7.

[14] H.R. Zeller, Die BBC-Flüssigkristallanzeigen—ein unbekanntes Stück Forschungsgeschichte, 25 Jahre ABB Forschungszentrum Baden-Dättwil (1992).

[15] R. Williams, Domains in liquid crystals, Journal of Physics and Chemistry 39 (July) (1963) 382–388.

[16] M. Schadt, W. Helfrich, Voltage-Dependent Optical Activity of a Twisted Nematic Liquid Crystal (TN-LCD), Phys. Rev. Lett. 27 (1971) 561.

[17] M. Schadt, M. Petrzilka, P. Gerber, A. Villiger, Polar Alkenyls: Physical Properties and Correlations with Molecular Structure of New Nematic Liquid Crystals, Molecular Crystal and Liquid Crystal 122 (1985) 241.

New Ventures

OBJECTIVES

The originator of a new venture starts with a fairly specific idea about the future business. Using four case studies, the objective is to show that there are no firm rules to be followed. Many changes in direction may become necessary. The use of common sense and persistence in driving one main idea may be two of the most important elements in making a new venture a success.

REQUIREMENTS FOR SUCCESS

"'For the *existing enterprise*,' the controlling word in the term entrepreneurial management is '*entrepreneurial*.' For the *new venture,* the controlling word is '*management*.'" (Peter Drucker, [1])

After the initial period of searching for the right idea that would lead to marketable products, the entrepreneur faces more unknowns than knowns. This can be good or it can be bad. Based on a few signs pointing to a later success, early decisions are taken—"common sense" is followed. While this may work sometimes, a few points need attention. My own experience, when involved in BBC Venture Capital Ltd., was the focus of one of our major venture capital partners in Boston on three points when reviewing a business plan:

- Value of the technology behind a new idea.
- Quality of the marketing information.
- Quality of the key people.

In more general terms, entrepreneurial management in a new venture requires

- A focus on the market.
- Financial foresight (e.g., planning for cash flow and capital needs ahead).
- Building a top management team early on.
- A clear decision by the founding entrepreneur about his or her own role and work.

209

MARKET FOCUS, FINANCIAL FORESIGHT, AND A TOP MANAGEMENT TEAM

In the beginning, the search for the right product or the right market may still be unfinished, the outcome open. Often, a successful venture ends up in a different area than originally envisaged. Products and applications may be different from the original ones, so may be the customers.

American Superconductors Corp. was initially entirely focused on developing products based on high-temperature superconductors and bringing them to the market. Only when the company redirected its focus and added power systems to its business did it reach real customers, and the revenues doubled within one year.

Nitinol Devices and Components started out with superelastic eyeglass frames and ended up with stents.

High-strength ceramics was the only constant in the development of Metoxit. Required skills, like precision grinding, had not been part of the original plan. Presintered slabs for machining CERCON® dental implants formed a new market, which evolved only after the success of CERCON®.

It is quite natural that the original vision of the entrepreneur is limited. The art is to be open to new opportunities and select those that will be important for success. It is impossible to do successful market research on something that is quite new, because we have to make assumptions about future customer acceptance. We can bring it to this conclusion: we have to work on the assumption that the new product may find customers in markets no one thought of. Without such an open-minded strategy, a market for future competitors may never be developed.

The entrepreneur often is in a dilemma. Venture capital companies focus early on profits, while the entrepreneur is more concerned with cash flow, capital needed for expansion, and control of the company's direction.

Very often, the founder is the person with good ideas but who has to learn that the art of survival requires the ability to build a team of capable people with a complementary skill set. Early on, this person also should think about his or her future role as the company evolves. I visited many venture companies, especially in the United States, where three years after the start of a venture, the founder became technical director, giving up the original function to a newly hired person with the necessary management and marketing skills to run the company.

DIVERSIFICATION WITHIN AN EXISTING COMPANY LEADING TO A NEW VENTURE

We will learn in this chapter that successful ventures in materials-based businesses may take 20 years. This may be a normal time period to establish a new material-based product, and it means that the culture of companies, their ownership, their core businesses, and the like will undergo significant changes. The following two case studies are quite typical of evolutionary change coming along with revolutionary ideas.

Typical changes in the culture of companies refer to their organization, the way they are managed. We will see that the freedom to make one's own decisions and the willingness to accept higher but still manageable risks results in a new culture, making these two cases possible. The most important success factor is the entrepreneur who has strong beliefs and visions, regardless of the environment about which the company is surrounded.

Case Study. Shape Memory Alloys Used in Stents [Personal Interviews with J. C. Palmatz, T. Duerig, D. Stöckel, and R. B. Zider]

Shape Memory Alloy Research at Brown Boveri

Nitinol is the acronym for the intermetallic-phase NiTi. William J. Buehler [2]), a researcher at the Naval Ordnance Laboratory in White Oak, Maryland, was the one to discover this shape-memory alloy in 1961. In the following years, lots of new alloys were developed and the underlying mechanisms studied. Three effects are representative for the unique properties of this class of materials, which immediately offered themselves for numerous potential applications:

- One-way effect.
- Two-way effect.
- Superelasticity.

Brown Boveri corporate research was continuously looking for new materials with interesting mechanical or physical properties, often as a function of temperature, with a potential to become useful in any existing products of the company. Previously, it had focused on high-temperature alloys, hard magnetic materials, contact materials, lanthanum-oxide cathodes, and materials with high damping capability. In 1975, it became aware for the first time of nitinol shape-memory alloys (SMA), which fitted the picture right away. After systematically going through all the existing lists of products, it found a potential application in LV (low-voltage) switchgear, which was produced in one of Brown Boveri's German companies. The product at that time used thermal bimetals, which was produced and supplied by Rau GmbH in Germany.

If it could have a switch that responds effectively to the relatively small thermal strains in bimetals, why not replace this material with a shape-memory alloy, which has a much larger strain? The business unit manager quickly bought into the new idea after an hour of discussion, and a small team of very competent scientists was assembled:

- Olivier Mercier, a Swiss physicist, who demonstrated that he could, in a very short time, find physical equations to describe the deformation behavior of shape-memory components of any geometry—rods, plates, coils.

- Keith Melton, one of a larger group of British metallurgists.
- Tom Duerig, a graduate from Carnegie Mellon University, who had originally joined Brown Boveri to work on high-temperature beta-titanium alloys.

Together, they started several research projects, such as shrink-fit applications for repair in power generating units, but they always kept their eyes open for new ideas. The research generated much internal interest in business units, and one of them even suggested creating a "Special Metals" division. Brown Boveri had no culture of setting up internal ventures, and no one had any experience how to do it. The best strategy at the time seemed to be delegating the work to internal and external consultants and using the research staff to provide information about the material and explain the potential applications. Since no one in corporate research was empowered to drive the process, most of the work was done by (inexperienced) consultants. There was no encouragement to work together as a team.

In July 1979, the business proposal was submitted to the board and turned down without prior discussion by the CTO and one of the top Swiss business managers.

In January 1980, the project was refocused on replacing thermal bimetal in LV switchgear. Soon, it became clear that NiTi was the wrong alloy for the intended application in LV switchgear, because the product had to perform the same way regardless if the external temperature was plus or minus $50°C$. Therefore, a new alloy had to be found, and this was to be CuAlNi. In a very short time, the alloy was developed, and Rau GmbH was contacted as a potential licensee. Dieter Stöckel, its chief metallurgist, became quite interested in the new materials.

In January 1981, Metallwerk Plansee in Austria was successfully contacted as a potential producer of the semifinished material, which a company like Rau would later process.

Since some of the required specifications could not be met by the alloy, corporate research decided to terminate the research project one year before the business unit followed with the same decision.

In December 1981, the main project was terminated. It had led to 15 patents and 21 scientific publications. Raychem, the leader in making shape-memory alloy shrink fits at the time, took a license on one of the patents related to this type of application.

Raychem then became like a sponge, sucking up the main specialists Melton, Duerig from Brown Boveri, and Stöckel from Rau.

Several learning points emerged from this experience:

- Shape memory alloys, because of their multiple ranges of properties, definitely had potential for uses by many companies, and it was a good decision to start research projects at Brown Boveri.
- Some of the best-qualified scientists teamed up with one of the top business units, and at an early stage, the issue of supply management had been addressed.

- The project was terminated at the right time; thus, it can serve as an example of a well-managed R&D project.
- Lack of experience on many levels of the organization led to the idea of contemplating a special metals business and employing inexperienced consultants to do the work.

While the scientists from Brown Boveri and Rau started work and established themselves at Raychem, another development was taking place in parallel—the development of stents.

The Story of Stent Development by Julio C. Palmaz

A common cause of heart attack and stroke is restricted blood flow caused by atherosclerosis, or clogged arteries. Millions who suffer from this condition have been able to avoid coronary bypass surgery and heart attack, stroke, and even premature death thanks to a revolutionary, implantable device, the balloon expandable stent, developed over nearly a decade by Julio C. Palmaz.

Prior exposure to state-of-the-art surgical treatments for cardiovascular disease made Julio C. Palmaz, an Argentinean, appreciate the potential of less-invasive techniques. In February 1978, at a meeting of the Society of Cardiovascular and Interventional Radiology in New Orleans, he listened to potential solutions presented by Andreas Gruntzig, using balloon angioplasty. Although Gruntzig was quite enthusiastic about the technology, he spent quite a bit of time explaining the limitations of the technique. From this one lecture, Julio learned of this amazing new way to open blocked vessels and that, often, the diseased vessels collapsed after balloon withdrawal by elastic recoil. Julio, faced with a problem, immediately had to think of a potential solution.

He bought copper wire and solder material at a local radio store and started to build first prototypes of the stent in his garage. The wire was woven in a crisscross mesh around a pencil with two rows of pins. Solder was used to fix the cross points to allow the mesh to retain shape. Once built, the mesh diameter was decreased by compressing it on progressively smaller wood dowels then was crimped by hand on a folded balloon. Inflation of the balloon inside a rubber tube left the stent affixed in place by friction after the balloon was removed. Although it looked fine from a geometrical point of view, it was clear to Julio that quite a few variables still had to be figured out, related to configuration of the mesh, thickness of the wire, expandability, and radial strength. Obviously, copper and solder were unacceptable implant materials, because they were not biocompatible. While thinking about how to make the mesh out of a single material, Julio got another idea. He noticed a piece of masonry metal mesh left in his garage by a carpenter. After cutting a piece and pressing it in a vice, the mesh openings became slots arranged in staggered fashion. The piece was rolled into a cylinder and the first model of a balloon expandable stent was born. To find the optimum geometry, Julio resorted to work with large cardboard models. This allowed him to vary the slot length, width, interval, expansion ratios, and so forth. It was clear that to scale down to the sizes needed for coronary applications would require sophisticated fabrication methods, with which he was not yet familiar. He was referred to a technician, Werner Schultz, who suggested using electromechanical discharge machining. In 1983, he began work as chief of angiography and special procedures in the radiology department at the University of Texas Health and Science Center at San Antonio (UTHSCSA). Dr. Stewart Reuter, chief of vascular procedures at UTHSCSA and a mentor to Palmaz, offered him facilities and research time to do the project.

University of Texas

The first stents were made by hand, of stainless steel wire soldered with silver, and successfully tried in an animal. The list of questions about the vascular stent and its potential kept growing, but the little research money available was drying up. The Department of Radiology had only limited funds that it could loan the project. The University of Texas Patent Committee rejected the application for a patent, and the federal government through the Veterans Administration rejected an application for funded research on the coronary stent. Julio teamed up with Richard Schatz, who was doing research on atherosclerotic baboons.

Philip Romano, an entrepreneur in restaurant chains, offered to put up $250,000 in exchange for a stake in the product. The trio, calling themselves the Expandable Graft Partnership, patented the stent technology in 1988 and presented it to a variety of large companies. Those firms included Boston Scientific, which passed on the technology, and Johnson & Johnson, which eventually licensed the stent technology for some $10 million plus royalties.

Clinical Experience

The first balloon-expandable stent placement in a human iliac artery was at the University of Freiburg, Germany, in May 1987, in a patient with a totally occluded left common iliac artery.

With Johnson & Johnson behind it and with an additional $100 million invested in its development, the Palmaz stent was approved for use in peripheral arteries in 1991, followed by approval for use in coronary arteries in 1994; Johnson & Johnson quickly captured 90% of the market for stents and bought the patent outright from Palmaz, Schatz, and Romano in 1998.

Competition naturally followed, with Boston Scientific, Medtronic, and Guidant sold their own stents, which resulted in a series of patent infringement lawsuits, most of which have been settled. Johnson & Johnson, however, saw its market share fall to less than 5% as competitors' new stent versions offered features cardiologists preferred.

A major breakthrough was the use of shape-memory alloys to manufacture stents. Figure 9.1 shows the use of superelasticity in shape-memory alloys in stents. Superelasticity allows passing a complex instrument through a cannula, returning to the deployed configuration once through.

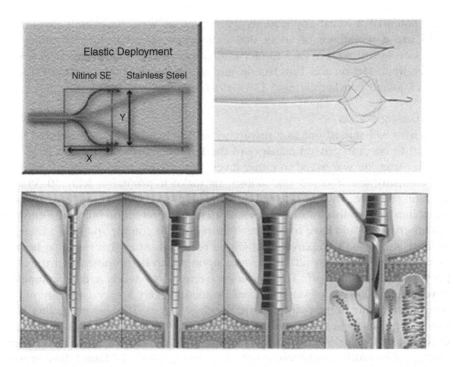

FIGURE 9.1 Superelastic shape-memory alloy stents and their deployment in a cannula.

Meanwhile, stent use had become increasingly common, with some 1.1 million implanted each year around the world, and used not only in coronary and peripheral arteries but also in organs such as the trachea, urethra, bile ducts, and colon.

Interestingly, the results with bare-metal stents are getting better, making cost considerations all the more important, as more patients and more lesions per patient are being treated with stents.

Several more learning points emerged from this experience:

- The idea was created because Julio Palmaz saw the problems of the previous technology as an opportunity to come up with a new idea.
- From the perspective of the early 1980s, the wide adoption of stents and percutaneous revascularization in general was impossible to predict.
- We have to accept early rejection of ideas and maintain the focus.

Nitinol Devices and Components

Raychem was one of the pioneers applying the one-way effect in shrink-fit applications; as mentioned earlier, other companies, among them Brown Boveri in Switzerland and Rau in Germany, became interested in using the two-way effect in SMAs (shape memory alloys) to replace thermal bimetals as actuators. Very early potential medical applications were cited as well. Raychem's division of SMAs had a workforce of about 200 employees, spending lots of money on R&D but losing money. Two former employees of the Brown Boveri Research Center, Keith Melton and Tom Duerig, and an employee of G. Rau GmbH, Dieter Stöckel, joined Raychem around 1983–1984.

Tom Duerig's View

Tom, who had started out as a researcher at Raychem, one day was asked if he wanted to become general manager of the Raychem Division. He accepted, but one day later, he was asked to either divest the division from Raychem, because the company felt it could not deliver, or make it profitable with 30–35% of operating income. The effort to sell the division proved fruitless, but management asked him to keep this information secret to save the retention bonus the employees otherwise would have to receive. Tom refused to comply and held a meeting in front of division's employees, which he recorded. After that meeting, he was asked by Raychem to lay off 100 people, upon which he proposed that the company should lay him off. The employees refused to work, which compelled Raychem to ask Tom to stay on. Tom decided to start a new company on his own, making and selling superelastic eyeglass frames. He was in contact with Bob Zider, from Beta Group, which owned the property rights, and John Fruth, chairman of Ocular Sciences Inc. He committed to make a $5 set after three months; John Fruth committed to sell the eyeglass frames. To found the company and do the incorporation, he chose the lawyer who had been asked to develop the plan for a management buyout. That business plan had been rejected by Raychem for two reasons: lack of experience with startups on the side of Tom and the potential embarrassment for Raychem should the plan succeed. He then used John Fruth's lawyer, which caused more conflict of interest. The learning from this is that, early on, one should choose a corporate lawyer with some equity interests, who also will sit on the board of the company.

The second issue was a certain lack of trust among the other partners, leading to more conflicts: Bob Zider, from the Beta Group with its own stock owners, had given an exclusive license to a Japanese company, which had cooperation with Marchon and Eschenbach in Germany to make and sell frames. Bob Zider then, in exchange for equity, offered a coexclusive manufacturing license to Tom but undermined it by selling exclusive marketing rights to Marchon. The learning from that conflict is that *early on we need to lock up the patent rights* in a clear and logical way.

To overcome the problem of low margins in the frame business in California, NDC set up a company in Cortina, Italy, which commissioned a manufacturing company to produce the frames for the paper company, which then sold it to Eschenbach. The original promise by Fruth to sell directly to the end customer had not been kept, and the investors decided not to further invest in NDC. Tom had long played with the idea to enter the medical market but was not supported by John Fruth and Bob Zider.

NDC was founded in July 1991. In January 1993, it started selling frames, which then were to become the base of revenue for NDC. Dieter and Alan joined at that time but had to assume large risks of their own: Alan would work as a consultant, paying both his and Dieter's salaries. After one year, Dieter was to have enough income from new sales of medical devices to get his own salary. For example, early on, ABB Sweden provided a contract for a shrink-fit application. At the end of 1993, the medical part became profitable, partially due to Dieter, who engineered stents via a $100,000 contract in KFZ Karlsruhe, subcontracting the sheet production. Since welding from sheet metal turned out to be a problem, stents were then made out of tubing by EUROFLEX, a company run in partnership with Rau. In early 1994, NDC employed 10 people, making just enough money to cover its overhead.

A huge debt accumulated and little support was given by the investors, since they did not want to enter the medical business and refused to supply enough capital to purchase manufacturing equipment. In mid-1994, the equity structure was messy. There was no capital to fund the medical business, and it became clear that NDC had to leave the frame business.

The equity structure consisted of

- $450,000 provided by private investors.
- The right to Beta Group's patent worth $1.1 million.
- $300,000 worth of stock owned by Zider, Fruth, and NDC.

The important financial number was cash flow. With a weekly cash drain of $30,000, the situation soon looked desperate, and it was decided to go into a debt of $750,000, funded from Fruth's own pocket in return for interest payment and debt converted to equity. The rest of the required funding was obtained by consulting work. The debt continued to increase to a maximum value of $1.1 million by early 1994, when it started to level off. By that time, it became clear that new capital had to be raised and some older problems had to be solved, primarily the messy patent and equity structure. It had turned out that the patent license was basically useless, because of the exclusive rights subsequently sold to Eschenbach. Eschenbach bought its exclusivity for $5 million, and Beta Group bought back the patent from NDC for $1.1 million. An independent lawyer was found, and the goal was to raise $2 million and partner with Teledyne Wah Chang, leading to a new equity structure:

- 20% new management.
- 40% previous owner.
- 40% Teledyne Wah Chang.

In addition, the previous owners had to retire the old debt. The previous owners had to accept medical applications and yield control of the board, giving up the chairman position. Since they would not accept this, Tom decided to quit but accepted a two-month period to make the deal happen (Dieter and Alan did not join him). The learning was this: *Keep your focus, even it means walking away from table once.*

Tom decided to invest in his own manufacturing of stents, making the company successful and profitable, and leading to an offer from Johnson & Johnson to buy NDC, which at that time had 40 employees. NDC did not make a price proposal but left it up to Johnson & Johnson to come up with one, resulting in a good deal for NDC.

The company quickly grew to an estimated annual turnover of $81 million, if it had been a stand-alone company and if one subtracted the cost of R&D assumed by Johnson & Johnson.

Here are some of the learning points:

1. While setting up a business plan is a good step to help you get organized, it is a certain failure if following it rigidly.

2. As a consequence, we always need to be flexible enough to change course.

3. The guiding principle in setting up NDC always was this: Corner the technology and people will come to you.

It is also sometimes interesting to recognize that each player in a business comes with his or her own thoughts, which may be not identical with the president's but may help broaden the picture. We therefore include the views of two of the other key persons: Dieter Stöckel, executive vice president, and Bob Zider, Beta Group (the venture capital company).

Dieter Stöckel's View

Stöckel's first exposure to shape memory was in the middle to late 1970s, when he was in charge of new technologies and materials at G. Rau GmbH in Pforzheim, Germany. In the early 1980s, the company had on the market a CuZnAl actuator for a ventilation system. Its efforts to develop a higher-temperature alloy on the basis of CuAlNi failed, despite the collaboration with Brown Boveri. It was also unsuccessful in purchasing NiTi raw material from Krupp in Germany and Raychem in the United States. At that time, these companies wanted to protect the technology and were interested in selling only finished products (like actuators), although these were not really developed yet. He visited Raychem the first time in 1979 to try to convince the firm that selling material to Rau would be a good strategy, but no agreement was reached. For a while, Rau worked with Furukawa material and gained some experience in wire drawing and forming. In 1984, it tried again to buy material from Raychem. Again, Raychem did not want to sell material but instead made him an offer to join Raychem as a process development manager. After many months of soul searching, he accepted.

While Raychem was very well equipped to produce fluid fittings (its bread-and-butter business) for military applications, there was no infrastructure for commercial products like wire- or sheet-based actuators. His group introduced multidie drawing, strand annealing, hot wire flattening, automatic spring winding, and other processes with which he was familiar from his activities at G. Rau. Eventually, he took over product management for commercial products, starting to sell springs to Mercedes and wire and a few components to the medical industry.

When Raychem's founder and chairman retired, the new management tried to get out of the shape-memory business and sell the Metals Division. It had no success but did spin off the Fluid Fittings Group. Tom left the company to start NDC, while Dieter stayed on and tried to market semifinished materials to the medical device industry with reasonable success. The new management at Raychem, however, became paranoid about liability in the medical market and made life difficult. In 1992, he left the company to join NDC.

In January 2003, Stöckel began as marketing director and member of the board of directors of NDC. His original role was to take care of the only customer, German eyeglass manufacturer Eschenbach, as well as to get the medical business going. His connections with a German medical device manufacturer generated $250,000 in year 1. NDC at that time had no means of manufacturing semifinished materials and basically bought and resold everything. One of the suppliers was G. Rau in Germany, which actually also produced the material that NDC would sell to the German customer. To streamline the process and the supply chain, a 50/50 joint venture with G. Rau was started in December 2003. G. Rau contributed the manufacturing expertise, while NDC provided the market knowledge and nitinol science. This joint venture, EUROFLEX GmbH, was chartered with selling all nitinol products produced either by G. Rau or NDC in Europe. The agreement with G. Rau also made NDC the exclusive distributor (reseller) of nitinol tubing produced by G. Rau outside Europe. Rau had started the development of the drawing process for nitinol tubing at NDC's request in early 1993 and presented the first marketable tubing about one year later. Around the same time, it built a relationship with a laser cutting company that had developed a process to cut stainless steel stents from tubing. In the 1994–1995 timeframe, NDC was the only company to offer nitinol stents laser cut from tubing.

Despite being quite successful in the medical market, the board (particularly Zider) considered the company a loser, because of lack of success in the eyeglass market (not possible within the framework of the license deals Beta had in place). Board meetings were always very frustrating, and the situation escalated to the point where Zider threatened to shut down the company and NDC's directors threatened to leave. In the end, Zider agreed to let the company look for a new investor. They found an investor in Teledyne Wah Chang. Tom and Dieter made a two-hour presentation of the company's business plan to the president of Wah Chang and walked away having basically sold 40% of the company for $2 million. The deal closed in a couple of months and life became easier. The new money was used to pay down the Beta debt and invest in laser cutting. The company now had a supportive board that included "real" business people with a lot of credibility. NDC quickly became the premier nitinol laser cutter, not only in the United States but worldwide. In Germany, the EUROFLEX joint venture entered into an agreement with a scientist from the Forschungszentrum Karlsruhe, who had developed a better cutting process; and in 1996, it founded EUROFLEX Schuessler GmbH, now named ADMEDES GmbH. EUROFLEX owns two thirds of ADMEDES, the balance is owned by the Forschungszentrum Karlsruhe scientist. ADMEDES today is the leading primary nitinol stent producer (not Johnson & Johnson) with 185 employees.

In early 1997, NDC was sold to Johnson & Johnson Cordis. Cordis became the worldwide market leader in self-expanding stents and vena cava filters within only a few years.

NDC had a few keys to becoming successful:

1. Profound understanding of the technology.

2. Genuine enthusiasm for the technology.

3. Willingness to partner with suppliers (Rau, Teledyne Wah Chang) and customers (Scimed, Medtronic).

4. Relationships built on trust and understanding.

5. Never taking yourself too seriously.

6. Luck (being at the right place with the right technology at the right time).

Bob Zider's View (Beta Group)

Applications of shape-memory alloys outside of shrink-fit applications appeared in the late 1970s.

In 1979, John Krummy invented the intravenous flow controller, into which Bob invested personally [3,4]. Further developments were the SMArt Clamp and Nanomuscles.

In 1983, the Flexon eyeglass frames were invented and developed into a business that lasted from 1985 to 1988. The product was offered to Raychem, but was turned down because it seemed to be too expensive. The key issues were related to the processing: hydrogen embrittlement, joining and forming.

With $1000/lb for NiTi, initial costs were too high. Material for $3-4 million was sold and eventually 80 million frames produced. The supplier of the material was a Japanese company, which later infringed on the patent. Manufacturing cost eventually was brought down to less than $20, and the frames could be sold for a profit.

Manufacturing licenses were given to a Japanese company, Marchon, and Eschenbach (Titanflex). Eschenbach maintained manufacturing rights in the United States and set up a plant with equipment from an earlier plant in Rhode Island.

When Tom started his own company making eyeglass frames, he soon brought down the costs to $18, while the Japanese manufacturer sold for $28. Tom became frustrated, because the Japanese company did not want to buy his glass frames, although they were better and cheaper. The Japanese held all the manufacturing licensees hostage. As a consequence, Tom soon ran out of money and the investors got tired. Mainly Dieter Stöckel, but also the others, felt that at least 10–20% of sales should come from medical products. A joint venture with Handy & Harman failed, whereupon Dieter set up a joint venture with G. Rau to manufacture tubing for stents. Tom found a company that knew laser cutting. It was important that the product have lots of value added to prevent it from becoming a commodity. Another technology needed was deburring. After three months of frustration, the polishing technique was sold to Boston Science for $3 million. Tom wanted to have an "open wallet" to further advance his business ideas.

The view of the investors was this:

- The original dream was a $70-million business in eyeglasses achieved by cost cutting, but this was prevented by the Japanese *licensee*.
- The original tubing business also failed, because it was becoming a commodity.
- It was always considered essential to also develop a medical business, but first the eyeglass business had to become profitable.

Then a new investor, Teledyne Wah Chang, was found, and the business objective became guide wires and tubing and stents.

Some important points emerged:

1. Materials science is only an enabling factor; the key is to have a proprietary application of a new material.

2. Typically ventures in materials fail, because materials often are 20 years away from an application.

Learning Points from this Case

To become successful in creating a venture you have to be in command of the technology.

To pursue your goals, you have to be totally unconstrained to readjust your course whenever this becomes necessary. Following this course takes a combination of enthusiasm, trust, and luck. Business plans may help you think ahead, but to follow them rigidly certainly leads to failure.

To develop a material-based new product can be a lengthy process. Make sure your application is proprietary and you do not lose sight of your vision.

Case Study. Metoxit: High-strength Ceramics in Biomedical Applications [Personal Interview with W. Rieger]

Objective

To move from a technology within an existing company through various stages of business development is a process with many unknowns, including the final target. This case study shows that a strictly rule-based approach relying on management theories is going to fail. The importance of using intuition, speed, and persistence is highlighted.

Metoxit

Metoxit is a medium-sized Swiss company focused on a large range of products using alumina- and zirconia-based materials. Figure 9.2 shows examples of dental applications (root posts, abutments, CAD/CAM blocks, implants, drilling tools). Figure 9.3 shows ceramic hip implants. The main materials used are alumina, yttria (TZP), and alumina-toughened zirconia (ATZ).

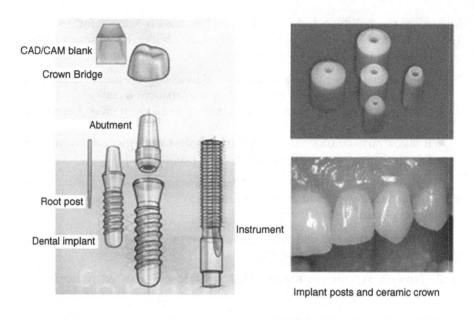

FIGURE 9.2 Example of ceramic products for dental applications.

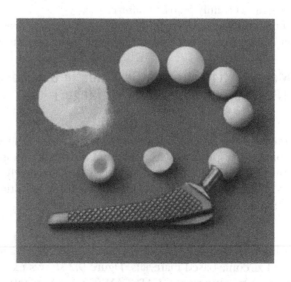

FIGURE 9.3 Hip implants, showing various processing steps.

As an example, the basic processing steps for the manufacturing of bioceramic hip joints in surgical-grade alumina and alumina-toughened zirconia are listed:

- Raw material formulation.
- Cold isostatic pressing.
- Green machining.
- Sintering in oxidizing atmosphere.
- Post-consolidation by HIP (hot isostatic pressing).
- Ball grinding and polishing.
- Cone grinding.
- Quality control.

From Innovation and Diversification to Startup in Alusuisse

In the early 1970s, the large aluminum company Alusuisse (now with the Alcan group) started building up a research group, focusing on materials, processes, and products with the goal of producing aluminum more efficiently and finding new markets for Al-based products. Some of the first projects focused on nonconsumable anodes for the electrolysis of alumina. Special electrically conducting oxide ceramics, based on SnO_2, were selected because of their conductivity at high temperature and resistance against aggressive melts.

Alusuisse R&D Strategy (1970–1985)

The project portfolio consisted of innovation projects directly related to the existing business and diversification projects (new technology, new market).

Innovation projects were related to the basic needs of the aluminum industry:

1. Nonconsumable anodes (based on oxide ceramics).

2. Direct reduction of Al_2O_3.

3. $AlCl_3$ electrolysis.

4. Direct casting of aluminum metal sheet (Caster 2).

5. Automotive applications of aluminum (e.g., body parts, suspension, clutch, brakes).

6. Magneto-hydrodynamic transport of liquid aluminum.

7. Thixotropic casting.

Diversification projects not connected to the basic needs of the aluminum industry included

1. High-purity gallium.

2. Gallium arsenide.

3. Wafers.

4. Packaging foils.

5. Catalysts.

6. Zirconia powder for the ceramics industry.

7. Industrial ceramics.

The big projects, 1 to 4, costing up to several hundred million Swiss francs were terminated when the company came into financial turbulence (1985). They were the pet projects of some of the company's top managers. Some of the careers of these top managers were affected by these developments.

The minor projects were initiated by forming a team of young, ambitious scientists, who received some (partially excellent) business training and a certain credit to pursue the project. Most of the minor projects were eventually terminated, when the company found out that *even smaller projects require a lot of attention, personnel, money, and commitment from the company.*

Most of the project leaders left the company, some of them frustrated, and started something different somewhere else (including becoming professors at the Swiss Federal Institutes of Technology in Zurich and in Lausanne). Some of the projects were transferred internally to production plants interested in the technology; some were sold to external, interested parties but generally did not continue to exist in the original way.

Creating Metoxit

Wolfhart Rieger had always been driven to develop a new business based on metal-oxide ceramics. In 1973, he came up with *Metoxit* as a protected brand name. As the leader of a new group, he began to develop commercial products based on Al_2O_3, entering the market in 1974. Since Alusuisse lacked the infrastructure of a ceramics manufacturing company, the idea came up to cooperate with an established ceramics company; and in 1978, Metoxit AG, a 50/50 joint venture company between Alusuisse and Tonwerke Thayngen, a nearby ceramics company interested in renewing its product line, was established. The initial goal was to set up small-scale production of the nonconsumable anodes and other products for aluminum metallurgy, following investment in processing equipment for the pressing and sintering of ceramics. As it turned out, Alusuisse was not really very much interested in ceramics for the aluminum industry once the company started its activity in 1980 and asked for new products and new markets. This led to the idea of applying technologies that had been developed aside from the main project: high-strength, high-purity alumina to be used for, among others, medical applications. The main idea behind it was to use the hot isostatic pressing process to postcompact sintered alumina parts, such as sphere blanks for hip implants. This was followed by the development and production of zirconia spheres by an equivalent process. Around 1985–1986, Tonwerke Thayngen was acquired by AG Ziegelwerke (AGZ), a family-owned brick and tile manufacturer in central Switzerland. AGZ then bought the remaining 50% of Metoxit AG from Alusuisse. The former operational activities of Tonwerke Thayngen were gradually abandoned and all remaining activities were merged into Metoxit AG. AGZ, in 1990, acquired Saphirwerke Industrial Products (SWIP), a company that specialized in high-precision grinding and polishing of hard materials, including balls and cylinders. SWIP and Metoxit had already been working together, supplying the grinding capacity for the biomedical activities.

Metoxit left the Alusuisse group in 1986, when Alusuisse decided to concentrate on mainstream activities. Metoxit eventually succeeded in bringing the original project idea (industrial ceramics) to a successful end.

Metoxit, in the beginning, involved *innovation and diversification* within an existing company but later became a startup (with limited funding). Looking back, this was probably the optimum that could happen.

Evolution of a Product from a Material

High-purity alumina was first developed in the early 1970s. In 1980, at the Alusuisse Research Center, the first experiments on HIP postcompaction of presintered high-purity alumina were started. It is quite natural, at this stage of materials development, to ask the question, What could it be good for? The original idea at the time was to use high-density alumina to make containers for nuclear waste storage. This idea was also propagated by Asea, the world's leading manufacturer of hot isostatic presses at the time. The Swiss National Organization for Storage of Nuclear Waste (NAGRA) was the driver for the project, which would allow depositing highly radioactive nuclear waste in alumina containers and seal and compact them by hot isostatic pressing. Since 40% of electric power in Switzerland came from nuclear power plants, if successful, this technology would open a huge market. The result of the research, however, was mainly of scientific value, since the application for nuclear waste deposition was abandoned due to safety and risk considerations. More important, it was shown that with high-purity, high-density (hot isostatic pressed) alumina, highly superior fracture toughness and crack growth behavior could be obtained, as a consequence of the elimination of defects and pores by the HIP process.

Thus, another project, using the high-density, high-purity alumina for medical applications, was becoming more successful. In 1983, the production of as-HIP alumina products, such as spheres for hip joints, was started. Soon competitors took up this new technology. Unfortunately, no patents could be filed due to the project with the state-run NAGRA organization. This increased the pressure to come up with even more-advanced ideas. In 1985 zirconia-TZP was added as another HIP material, and through the grinding know-how and diligence of the soon to become sister company SWIP, an unbelievably precise surface and shape of the spherical balls could be obtained. The typical deviation of the spherical shape in any direction is 0.1 μm, the surface roughness factor Ra is 0.02 μm. Thus, after the acquisition of SWIP, fully integrated production from powder to finished product became reality. From 1989 to 1991, an application at the U.S. Food and Drug Administration led to the issue of master files for alumina and zirconia for orthopedics in the United States, and similar permits were issued in Europe.

In 1991, Metoxit was the first ceramic company to enter the dental market with zirconia-TZP products. The first products in this new market with what became termed *zirconia bioceramics* were posts for tooth roots. Its unique selling position at the time was its worldwide leadership in high-strength HIP zirconia and its market access in Switzerland and Germany. The competitors either ignored the challenge, because they were focused on other materials, or had more limited market access. In 1993, manufacturing of sintered blocks of zirconia for CAD/CAM-machining of crowns or bridges, to be performed by the dental technician, were added to the product portfolio, as equipment for precision machining of such blocks had become available (see also the case study CERCON® in Chapter 1). Nevertheless, Metoxit continued to stick to its conservative, top-quality manufacturing technique: shape, sinter, grind, and polish. In 2001, the orthopedic market for zirconia, and Metoxit, was severely harmed by the mishap of a French competitor, who subsequently retreated from the bioceramics market and stopped all activities in orthopedics and dentistry. The dental market was not affected, but there was new pressure for more innovation. Therefore, in 2001, the new material ATZ, with a strength of 2000 MPa, was added to the program and later used as the first oxide-ceramics drilling and grinding tool for dental applications. The combination of several innovations made it possible to offer a whole spectrum of dental blocks for CAD-CAM (HIP, porous, unsintered) by a fully automated process.

Figure 9.4 summarizes the history of Metoxit and the evolution of its technologies.

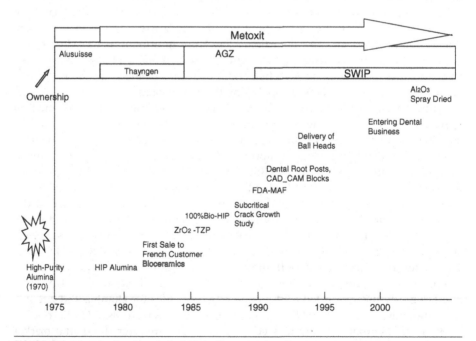

FIGURE 9.4 Historic development of Metoxit.

Guiding Principles, Role of External Funding

Metoxit's founder has been driven by a few, but very simple guiding principles:

1. Carefully scan the business world around you.

2. Select a few (not many!) project ideas that promise a unique selling position.

3. Start to realize your idea.

4. Build up enough self-confidence to be able to defend your idea against the "silly" questions of bankers, consultants, and business managers with only an MBA background.

5. Try to survive business plans and do not believe in them.

6. Stick to your idea.

7. External funding may slow you down.

Although external funding should not be totally discarded, it can slow you down when trying to build a business. The process of getting external funding costs lots of time and creates much idle activity, due to the bureaucracy from the side of funding agencies.

Seven Basic Ideas and Learning Experience

Idea 1. Alusuisse needs production facilities for larger-than-lab-size quantities of special ceramics to be used in aluminum production technology—Rating: Quite good idea.

Idea 2. Coalition or cooperation with a nearby, old-fashioned ceramics company looking for a new leg (Tonwerke Thayngen) to establish prototype manufacturing facility (50/50% ownership)—Rating: Not such a good idea.

Idea 3. Manager of new company, Metoxit AG, believes in (a) pronounced interest of 50% owner Alusuisse in capability and capacity of new company—Rating: Bad idea!—(b) impact of 50% owner TWT to aid startup in marketing and technology—Rating: Very bad idea!

Idea 4. Manager of Metoxit believes in *"Help yourself so help you god!"* and, successfully, scans the various fields of applications of ceramic components, eliminating (among others) automotive products, water faucets, paper mill machinery, textile thread guides, mill linings, and grinding media but concentrates on (among others) cylindrical parts, spherical parts (balls, spheres), medical applications.—Rating: Very good idea.

Idea 5. Manager of Metoxit believes in idea of bypassing Alusuisse mainstream career steps by creating something *small, new, not mainstream but with highly interesting long-term perspectives*—Rating: Short term, terribly bad idea; long term, best idea in manager's life!

Idea 6. Manager created name Metoxit in 1973 because he believed it was self-explanatory (METal OXIde)—Rating: This turned out to be quite a good idea; but it will take 25–30 years to really make the name known and respected worldwide.

Idea 7. Management and board of Metoxit's new owners (since 1986, separated from Alusuisse and its burdens) acknowledge the idea that only *fully integrated production, commanding all necessary technological steps, including ball polishing and high-precision grinding, leading to the takeover of SWIP(1990) can be successful.*—Rating: Excellent idea!

STARTING A NEW VENTURE FROM A NEW TECHNOLOGY

While the two previous cases have been very challenging, not only technically but because of nonanticipated changes in the environment, they had the benefit of going through a learning phase within an established company. In many more cases, the founder is focused from the very beginning on creating the venture. The founder may come from a university or research organization. The next case describes the rapid buildup of a new venture from a university/science park environment.

Case Study. Amroy: Building a Business Around Carbon Nanotubes (Personal Interview with P. Keinänen)

Carbon nanotubes (CNT) were first produced by Sumijo Iijima in 1991 [5], although two Russian scientists, L. V. Radushkevich and V. M. Lukyanovich [6] reported about them in 1952. These giant molecules are among the most important materials innovations due to their exceptionally high mechanical strength and electrical and thermal conductivity. There is a continuous need for stronger and lighter materials.

Although it seemed to be an obvious next step, CNTs and other nanosized additives have not been used successfully as strengthening components in a plastic matrix due to the poor bonding. For the best results nanosize additives must be chemically compatible with the surrounding matrix.

Three researchers at Nanolab Systems Oy, a holding company to encourage new product development and spin-offs at the NanoScience Center in Jyväskyla, Finland—Pasi Keinänen, Jorma Virtanen, and Mikko Tilli—were working together. In 2003, the academic research began. In 2004, a new research idea was born. While trying to develop sensors for detection of explosives, they tried to find a way to disperse CNTs. By sheer accident, they discovered a new technology: how to create radicals at the end of CNT molecules by ultrasonic vibration. Other methods like mechanical or optical cutting of the CNTs could be used as well to break up the CNTs, creating highly reactive radicals. This made it possible for epoxide and other functional groups to chemically bond to the modified CNTs (Figure 9.5).

This fusion structure is now known as Hybtonite® epoxy. It is the only composite resin where the carbon nanotubes are joined with the epoxy molecules

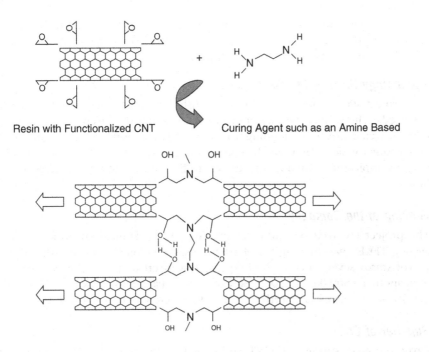

Resin with Functionalized CNT   Curing Agent such as an Amine Based

FIGURE 9.5 Creation of CNT-reinforced epoxy by simultaneous application of ultrasound and a curing agent.

through a covalent bond forming a cross-linked three-dimensional carbon atom lattice. This novel hybrid material can outperform other additive-based carbon nanotube resins by pure chemical power and the nature of covalent bonding. The Hybtonite® epoxy resin is suitable for high-performance coatings and all composite processes from injection to prepregs due to the controlled viscosity and performance out of the covalent bonds.

Since the team had no epoxy production facilities, it looked for an established partner in the field and found Kimmo Kaila in fall 2005, which had come from Shell and founded SBT (Specialty Binder Timber) Oy in 2002. SBT Oy already had an established customer base of 200. Amroy Oy was founded and the patents sold from Nanolab Systems to Amroy for €80,000. The first products were hockey sticks, developed in 2005 for Montreal Sports, owned by a NHL player, who had his own production line. Lab tests at the University of Tampere showed 70% better energy absorption than the established product. Very early, other sports equipment, such as cross-country skis and lightweight boots for ski jumping, was tested and brought successfully to the market. The range of other applications is huge, covering all fields in the automotive, aerospace, and construction industry, where high-strength, lightweight materials are needed. One of the potentially largest applications is for wind turbine blades. The blade would be 61.5 m long and weigh 18 tons with 10 tons of resin. The motivation to switch to epoxy resin from polystirol is the emission of styrene, which is carcogenic. At present, the material is in the test phase for this application.

Technology Push or Market Pull

The project started out as a technology push, but soon was focused on real market needs. The main initial drivers were sensors, but the discovery of new methods to create covalent bonding to other materials opened the field of new applications widely. Lightweight composite materials had long been established in many applications, so now the art was to find those that saw the largest benefit first.

Funding of the Company

The project received only indirect funding, as the Finnish government funding agency, TEKES, was funding Nanolab Systems, the holding company. After the creation of Amroy, several focused TEKES projects were created. Private investors contributed another €500,000. In 2007, 80% of the company was owned by the founders, 20% by outside investors. Further €1.2 million were raised in summer 2007.

Supplier of CNT

Bayer, the main supplier of CNT, succeeded in dropping the price per kilogram from €1 million in 2002 to €150 in 2007, and the present production capacity is at 60 tons/year.

Production and Sales Growth

Although the company is still very small, production grew very rapidly, mainly because of partnering with SBT Oy.

2005   5 tons.
2006   200 tons (including the volume of SBT Oy after fusion).
2007   350 tons.

The company had annual sales of €2 million, seven employees, and a 4000-ton production capacity. The average annual growth rate had been at 100%, keeping the net profit zero.

Growth Strategy and Main Risks

All companies, weak and strong, at this embryonic stage of development, emphasize the development of a distinctive competency and an associated business model. During this stage, investment needs are great. Companies require large amounts of capital to build up new competencies, such as production, sales and marketing, and services. A company's success depends on its ability to demonstrate a distinct competency to attract outside investors. Entering a growth phase, the task facing a company is to consolidate its relative competitive position in a rapidly expanding market. Since other companies may enter the market and catch up with the first movers, the first movers often require successive waves of capital infusion to maintain the momentum generated by their success in the embryonic stage. Companies unable to find all the additional resources may engage in a market concentration strategy to consolidate their position. Partnering with other companies, such as SBT Oy, is one way to surpass the need to invest in building up all the competencies needed.

The case study on LCDs (Chapter 8) shows that, even with a strong patent position, new companies may enter the playing field and leave the first mover behind, because they may have other skills the first mover company lacks.

Patent Situation and Competing Technologies

Although the CNT treatment and bonding process to the epoxy matrix has been patented, other companies, such as Zyrex in the United States, have come up with a competing technology, using surfactants; and due to the many variables in this area, they will not be the only ones. China, India, and North America have to receive special attention as potentially huge suppliers and markets.

Learning Points

1. Amroy, although the result of a technology push, evolved in the right environment. Finland is a leader in Europe in terms of innovation. Although the government is funding projects, it delegates lots of decision power to companies, helping them quickly find their own path.

2. The team of the founders of Amroy brought a large combination of skill levels: many years of research in nanotechnology, some early experience in creating new companies.

3. Recognizing that growth is important but it also requires lots of capital, early on, the team made an important decision to partner with an existing company that had both the skills in epoxy manufacturing and an established customer base.

4. Securing a good patent base is mandatory.

5. Finding the right growth strategy is a huge challenge, and many approaches are feasible.

Case Study. Day4Energy: From Auctioning Off a New Technology to Setting Up In-house Manufacturing [Personal Interviews with J. MacDonald in 2003 and 2006]

The field of photovoltaics, like wind power, has grown steadily over the years and undoubtedly will play an increasing role in energy production in the future. This has not been obvious for a long time, and here we can credit government-funded research as a reason for continued progress in materials development. Not only did governments contribute to the further development of the technology, but some countries like Germany and Japan provided financial incentives to invest in this technology by buying back excess power into the electric power grid.

Photovoltaic Technology

Photovoltaic (PV) cells are semiconductor devices that convert photons into electricity. They can convert both sunlight and light from artificial sources. A typical silicon PV cell is composed of a thin wafer consisting of a very thin layer of phosphorus-doped (n-type) silicon atop a thicker layer of boron-doped (p-type) silicon. An electrical field is created where these two materials are in contact, called the *p-n junction*, and exposed to light. The cells produce 0.5–0.6 Vdc under open-circuit (oc), no-load conditions. Both current and power output depends on the cell's efficiency, its surface area, and the intensity of the light. For example, a PV cell with a surface area of 160 cm^2 produces about 2 W peak power in full sunlight, but only 0.8 W with 40% intensity of sunlight.

A multitude of materials have been developed as potential PV cells. They differ in energy conversion efficiency and electricity generating cost. Energy conversion efficiency for commercially available mc (multicrystalline)-Si cells is 14–16%. Much higher efficiencies, up to 40%, can be reached with multijunction cells, but they may cost 100 times as much as 8%-efficient amorphous Si in mass production, while delivering only 4 times the electrical power. Electricity generating costs are still high, $0.30–0.60/kWh, compared with $0.04–0.05/kWh for commonly available power-generating technologies.

Since power is proportional to the intensity of sunlight, so-called concentrators (lenses or mirrors) are used to focus the sunlight on a small area of PV cells. Since the sun moves continuously, single- or dual-axis tracking is used to improve the performance. The main advantage of using concentrators is to reduce the amount of Si, which is both expensive and increasingly in short supply. The disadvantage is still related to the fairly high costs of the focusing, tracking, and cooling equipment.

Figure 9.6 shows the standard multistep manufacturing process, consisting of first creating the doped layers and finally applying the contact material, which draws the current from the cell and moves it into the current conductor.

A series of such cells is laid out on a plate to lead to a module; several modules are connected into an array. So, how can we take serious steps forward to increase efficiency and reduce the cost of electricity? Shall we start from the still too expensive high-efficiency cells and find new ways to reduce the cost or stay with the well-understood Si cells and try to cut costs and improve efficiency? Day4Energy chose the second approach, but had to readjust its strategy, as the company evolved.

History of Day4Energy

John MacDonald cofounded the company. A graduate from MIT (1964), he became the cofounder and CEO of MacDonald-Dettwiler, Canada's leading space-robotics firm, which he led for 30 years. As a member of the business

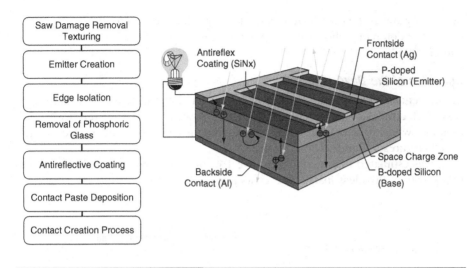

FIGURE 9.6 Front-side metallization with screen printing.

advisory council of APEC (Asia-Pacific Economic Corporation), he attended a meeting in Moscow, May 2001. His son suggested that he meet Professor Leonid Rubin, whose son lived in Vancouver and was his friend. After visiting several companies and research labs, he was very impressed by the many ideas and knowledge of Leonid. After a few days Leonid asked John, "Which one of the many ideas could we build a company around?" John's answer, "It's obvious— your solar energy project. If you can do what you say you can do, and I see no technical reason why you can't, there is a market." Day4Energy was founded in 2001.

My first contact with the company was in 2003. The company consisted of a small but very efficient lab, employing about five to seven persons. The proprietary technology is based on the use of a newly developed low-melting alloy with high electrical conductivity, which could replace the current collection leads on the front end and save about 70% of the cost of the expensive Ag-paste used then. More important, the efficiency could also be increased. A pie chart showed the distribution of PV manufacturers worldwide. There were about four or five major ones and a large number of small ones. Since all use basically the same process to produce the cell, the cost-saving new technology could be used by all of them. The vision of the first product was a concentrator, making systems of 500 kW possible at a competitive cost—"enough to power a supermarket."

A probably optimistic cost calculation gave an estimate of 8 cents/kWh for the cost of electricity, which would compare favorably with well-established power-generating technologies. A quick tour through the lab showed that all the processing steps had been demonstrated. The business concept at the time was a "no brainer": "The advantage of this new technology is so obvious, and it can be used by any of the established manufacturers, that we plan to offer the technology to the company who wants to pay the most for it—we will auction it by the end of 2004."

Specifics of the Day4 Electrode

In the regular cell current collection is restricted to relatively centralized Ag-pasted electrodes. The Day 4 Electrode is a low resistance current collection method, where the collectors are distributed densely over the surface of the Si cell (Figures 9.7 and 9.8).

It was estimated that mc-PV cells give 15% more power per square meter and require about 20% less installation costs per square meter.

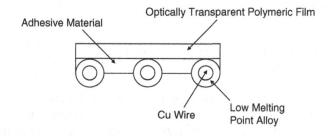

FIGURE 9.7 Concept of the Day4 electrode. Wires are fixed on one or two Cu busbars, which collect the current from the cell via the electrode wires.

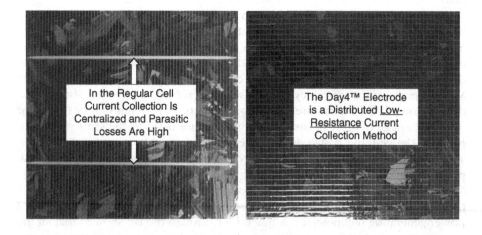

FIGURE 9.8 Centralized current collection (left), decentralized current collection (right).

Further Development

A second visit, in summer 2006, showed that the company was going strong, although the original objective had not been achieved. The market was not ready for solar concentrators, and the cost of concentrators turned out to be higher than expected, so the company turned to lower-cost, higher-efficiency flat panels. The "auction" hadn't taken place, but the company was in a rapid buildup stage for in-house manufacturing and had reached an agreement with the largest German PV company, Q-Cells, to apply the Day4Energy technology to its cells. Subsequently, the company diversified its cell supply strategy by signing agreements with two Taiwanese cell manufacturers in addition to Q-Cells.

The goal of sun concentration is to reduce the amount of PV material per watt generated, thereby reducing the cost per watt. Conventional multicrystalline Si PV cells do not work efficiently under concentrated light. The advantage of Day4's technology is that it permits a conventional PV cell to function properly under concentrated illumination. The Day4 concentrator concept uses industrial cells and does not require more than 10 suns of concentration. This lower concentration also means that lower-cost one-axis tracking is sufficient; however, at this point in time, the cost of a concentrator system, once commercialized, was found to be too high due to the increase in cost represented by the need for an optical system, a tracking system, and a heat dissipation system. These costs at the commercial level were initially underestimated.

The present strategy is clearly focused on rapid growth. This required new capital, which led to a multiphase financing strategy beginning with venture capital/angel financing in 2004–2006, through institutional financing in 2006–2007, culminating in a $100 million independent public offering in December 2007.

In early 2007, the company set up a manufacturing capacity of 12 MW to be followed by a further expansion to 47 MW by early 2008. As sales developed, the company found that the vast majority of its 2008 sales were in Europe. Therefore, in late 2008, the company outsourced a further 50 MW of production to a contract manufacturer in Europe, which is expected to come on line in mid-2009. The expansion is driven by fully utilized present manufacturing capacities and an order backlog stretching into 2010. Employment is now (January 2009) 174 persons.

Learning Experiences

John MacDonald sees the learning experiences so far in the following way. In start-ups, changes are normal and we always have to be ready to readjust the strategy. It is important to recognize the need for readjustment early rather than follow the initial strategy at all cost. If you like your technology, then you still have to convince your customers and plan for the time required to go through this learning period. Early on, work with potential suppliers, such as Q-Cells. Be ready to recognize the special culture of venture capital companies, as they are mostly driven by short-term returns, while the buildup of a new business often follows different rules.

SUMMARY

New ventures may start as spin-offs from established companies, which do not see the idea fitting into the core business strategy. Ventures can also start from zero. Both approaches are described by two case studies each.

NDC started out with a research project on shape-memory alloys at BBC corporate research. The founder of NDC underwent many learning experiences but never lost sight of his motivation to corner the technology.

Metoxit, in a similar starting phase at a large Swiss aluminum company's new research center, was driven by the idea to come up with stronger, better ceramic materials and find a market for biomedical applications. Becoming successful needs more than mastering tools and financial support: Intuition and common sense are more important to ultimate success.

Amroy was created in the highly creative environment of a Finnish science park. The idea came accidentally but was picked up immediately. The value chain was quickly completed by teaming up with a major manufacturing company, and an amazing range of applications were brought to market in a very short time period.

Day4Energy was the result of an experienced Canadian entrepreneur meeting a highly innovative Russian professor. Like in many other startups, the original business model had to be dramatically changed. The ability to make changes quickly and in the right direction is the common denominator in these four successful startups.

REFERENCES

[1] P.F. Drucker, Innovation and Entrepreneurship, HarperCollins Publishing, New York, 1999.

[2] W.J. Buehler, R.C. Wiley, J.V. Gilfrich, Effect of Low-Temperature Phase Changes on Mechanical Properties of Alloys near Composition TiNi, Journal of Applied Physics 34 (5) (1963) 1475.

[3] R.B. Zider, J.F. Krumme, Eyeglass Frame Including Shape Memory Elements, U.S. patent 4,772,112 (1988).

[4] R.B. Zider, J.F. Krumme, Eyeglass Frame Including Shape Memory Elements, U.S. patent 4,896,955 (1990).

[5] S. Iijima, Helical microtubules of graphite carbon, Nature 354 (1991) 56–58.

[6] L.V. Radushkevich, V. Lukyanovich. M., О Структуре Углерода, Образующегося При Термическом Разложении Окиси Углерода На Железном Контакте, [Soviet] Journal of Physics and Chemistry 26 (1952) 88–95.

The Human Factor in Management

OBJECTIVES

This chapter begins with an analogy between mountain climbing and management. The objective is to understand several of the similarities.

The objectives of the description of psychometric instruments is to make you familiar with the value to

- Find your preferences and traits, linked to your personality.
- Measure your skills and abilities, which you can work on to improve.

You will understand important characteristics that define a good manager:

- The role of intuition and common sense.
- How to manage people well and delegate.
- How to make a good choice of consultants.

LEADING OR BEING LED: ANALOGY BETWEEN MOUNTAIN CLIMBING AND MANAGEMENT

Let us start with an analogy between mountain climbing and leadership in teams to get a first sense of what it takes to become a leader, mainly based on my personal experience.

My first exposure to climbing came at the age of 20, when I decided on taking a rock-climbing course. The course started by learning how to tie a rope around yourself. About 20 m above ground, at the end of a rockface, the teacher swung a rope around a tree and held on to it to belay: If I slipped and fell, the rope would tighten and keep me safe. Very early, I felt really good when the teacher was making positive remarks about my climbing style, and my confidence increased rapidly. Later on, we climbed greater lengths, which required at least two persons connected by a rope. The first one was the leader, the second one the follower.

237

Soon we changed the leadership—the follower had to pass the leader (or, leadership was delegated to the follower), thus becoming a leader and establishing a secure position to allow the leader to follow and pass. Communication was very important. Very clear commands agreed on before the start of climbing, such as "secured—climb," and how much rope was left to warn the follower ahead of time when to look out for a place to secure himself or herself. The teacher remained positive in commenting on my climbing attempts. It only took three or four lengthy climbs to distinguish between the different feelings I had when I was leading and when I was being led. Being the leader automatically meant more responsibility—toward myself, because I had to decide which were the right steps to take, and toward the person following me, because now I was also responsible for that person's safety.

My job then brought me to Switzerland, known for many of the highest peaks in Europe. I started to climb many of these mountains, but with one exception I found partners only when I agreed to lead, some of them coming from countries as far away as Scandinavia, the United States, and Australia. Many of them had never climbed a mountain or been roped together. So, now, I became the teacher and the one who had to use encouraging words to help the beginners. In a few instances, my partner would show moments of fear, and I had to make a decision: Either I could succeed in making that person feel comfortable and safe and we would continue or I had to turn back. This happened a couple of times, but I never felt sorry, although this also meant that we missed making it to the summit. The experience of leading others in climbing also gave me the confidence to organize my own hiking trips in faraway countries, and I quickly learned that they were more interesting than organized tours: On an organized tour you are being led, while in a self-organized tour you are the leader.

Subconsciously, I always drew analogies of being roped together and leading a team, and the success of climbing gave me the confidence to lead and manage a team. As John Sims [1] said, "People have to know they can come to you with a serious issue and trust you to help them resolve it. I have used the analogy of being roped together as a team, the constant cycle of trusting while climbing on belay then being responsible as the anchor, numerous times."

We can learn much from other, more established climbers. Dan Mazur [2], leading a small group of clients to the summit of Mt. Everest in 2006 decided to abandon the summit and rescue the climber Lincoln Hall, who had been left behind as "dead" by his climbing partners 200 m below the summit. His clients did not object, although they had paid a large amount of money for this climbing expedition.

A leader has to use and cultivate several abilities:

- A willingness to instill an appreciation for risk in an uncertain environment.
- Building of trust in the climber ahead of you and responsibility for the person behind you.
- Effective communication—it has to be clear and consistent.

- Teaching the other person during difficult moments.
- Balancing the demands of results and relationships in the team.
- Not everyone wants to lead in climbing—not everyone wants or can be an effective leader in a business environment. *People are different from each other.*

A leader has to know what it takes to bring others to the top; and the followers on the team should know who is helping them to reach their goal. We soon learn that some people are afraid of steep hills and would never dare learn climbing, but we have to accept them like anyone else.

If we move away from this example of climbing and try to generalize and apply the same thinking to team management and leadership, we quickly appreciate that each person has abilities that can and should be enhanced and every person has a distinct personality.

TOOLS TO MEASURE OUR CAPABILITIES AND LEARN ABOUT OUR PREFERENCES (PSYCHOMETRIC INSTRUMENTS)

The field of human psychology is too vast and complex to go into too much detail, but it is helpful to know that many aspects of the human mind can be measured by so-called psychometric instruments. These instruments can be broadly divided into three categories [3] (Figure 10.1).

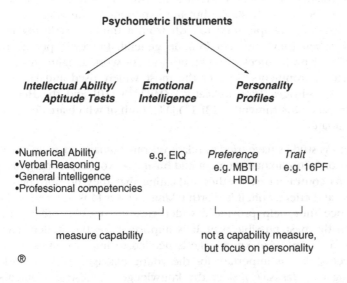

FIGURE 10.1 The three types of psychometric instruments.

- **Intellectual ability** (IQ) tests have been very common, but their correlation to management performance was seriously questioned when the topic of emotional intelligence was first introduced by W. L. Payne [4] and P. Salovey and J. D. Mayer [5] and promoted strongly by Daniel Goleman [6,7,8].
- **Emotional intelligence** describes abilities distinct from, but complementary to, academic intelligence, the purely cognitive capacities measured by IQ (intelligence quotient). Studies (e.g., [9]) show that little relationship between emotional and academic intelligence was found. Many people who are book smart but lack emotional intelligence end up working for people who have lower IQs but are superior in emotional intelligence skills. Simply put, there are 2 kinds of jobs that exist in this world:
  - ◻ Jobs that require a lot of analytical skills and know-how, but with limited or no interaction with other people. Typical jobs would be in research, mathematics, engineering science, economics, and the like.
  - ◻ Jobs that require a lot of until then unknown and now newly acquired skills with an intense interaction with other people. Examples are leaders and managers.
  - ◻ The first call for technical skills and competencies (IQ), and the second call for mainly soft inter- and intrapersonal skills (EQ or emotional quality).
- **Personality profiles** show the different ways people think and communicate. The diagnostic instruments used to differentiate between various personalities do not measure ability or intelligence and do not predict performance. Personality profiles measure preferences. This means they tend to be relatively stable, but life experiences can gradually affect them. For example, people tend to drift from a thinking style toward a feeling style when they have children. In general, however, people retain their dominant preferences throughout different work assignments and different social environments. Two of the most widely used and known tools are the Myers-Briggs Type Indicator® (MBTI®) [10] and the Herrmann Brain Dominance Instrument® (HBDI®) [11], both of which are described in more detail later.

Numerous studies have tried to relate various human factors to each other and to success or failure in management and business. Not surprisingly, many of these correlations contradict each other, indicating that they do not completely combine cause and effect. Still, it is worth taking a closer look at some of these instruments, since they help provide a wider view of oneself, others, and how to interact in the best possible way. It is important for the student and first-time employee to know what knowledge is needed to improve oneself and become more successful; it is important for the young manager, who is burdened with challenging new tasks. Consider the knowledge of these instruments more as an eye-opener than a foolproof recipe for becoming a successful manager.

Emotional Intelligence (EI)

Much of this section is based on Daniel Goleman's books *Emotional Intelligence* [6,8]). He defined EI as the capacity for recognizing our own feelings and those of others, motivating ourselves, and managing emotions well in ourselves and in our relationships. Success actually depends on a set of emotional and social competencies that builds on EI. He organizes these competencies into the five, and later four, *"Characteristics and Abilities of Emotional Intelligence Containing 21 Categories"*:

1. **Self-awareness** (Emotional Self-Awareness, Accurate Self-Assessment, Self-Confidence)
2. **Self-management** (Emotional Self-Control, Transparency/Trustworthiness, Adaptability, Achievement Orientation, Initiative, Optimism, Conscientious-ness)
3. **Social Awareness** (Empathy, Organizational Awareness, Service Orientation)
4. **Relationship Management** (Inspirational Leadership, Influence, Developing Others, Change Catalyst, Conflict Management, Building Bonds, Teamwork and Collaboration, Communication)

Emotional intelligence can be compared to a mirror in which we see ourselves or others. It is useful as a tool for project leaders or members in a project team who have no special education to correctly interpret group-dynamic processes.

It is easy to understand why the inclusion of emotional intelligence in training programs can help employees cooperate better and motivate more, thereby increasing performance.

There are a few benefits of emotional intelligence which should be mentioned:

- The importance of emotional intelligence increases, the higher one moves up in the level of management in the organization
- *Pattern recognition*, the *"big-picture" thinking*, helps leaders to identify the right trends from the huge amount of information around them and to think strategically far into the future. With this one exception, intellectual or technical superiority played no role in leadership success.
- Emotional competence makes a crucial difference between mediocre leaders and the best—most of their *success in leadership* was connected to emotional intelligence.

Emotional intelligence is thus an important element contributing to successful management.

Our emotional intelligence determines the potential for learning the practical skills based on its five elements (self-awareness, motivation, self-regulation, empathy, and adeptness in relationships). Our emotional competence shows how much of that potential we translate into on-the-job capabilities.

Emotional intelligence defines the maximum potential of skills that are based on the 4 elements (self-awareness, self-management, social awareness, relationship management), whereas *emotional competence* shows how much of that potential has been reached.

The Emotional Competence Framework

Goleman [6,8] put all competencies, personal and social, together in the Emotional Competence Framework. A detailed description of the content of the various skills is presented after a brief introduction.

Depending on the level of emotional competence, people find themselves either constructive or obstructive and inhibiting. In dealing with inhibitors, the core competence to be developed is self-awareness: recognizing our emotions and their effects on others. Once we understand the content of the various competencies, it becomes almost common sense how to deal with a deficiency.

A typical example of an inhibitor would be an aggressive person. What such a person would have to develop is empathy and social skills. *Empathy* means the ability to understand others and sense other's feelings; better social skills helps the person work with others toward common goals without using aggressive tactics.

Here are more detailed definitions of the various emotional competencies by Goleman. They should be read carefully before undergoing a self-assessment.

Self-Awareness

Self-awareness consists of the competencies *emotional self-awareness, accurate self-assessment, and self-confidence.*

Emotional self-awareness is the ability to recognize one's emotions and their effects. People with this competence know what they feel and why. They recognize how their feelings affect their performance and have a guiding awareness of their values and goals. If we lack the ability to recognize the correlation between emotions and what we are doing, we may be sidetracked by emotions we loose control of.

Accurate self-assessment helps us to be aware of our strengths and weaknesses. People with this competence can reflect and learn from experience. They are open to candid feedback and willing to continuously learn and develop further. They often show a sense of humour and perspective about themselves.

Lacking this competence, people often might show blind ambition (appearing right at all costs), set themselves unrealistic, overly ambitious goals, work hard compulsively at the expense of everything else; they seek power for his or her own interests rather than for the organization's. They always look for recognition at the expense of others and are mainly concerned with their prestige. Because they consider themselves perfect they reject even realistic criticism by others. Worst, they micromanage and take over instead of delegating; they come across as abrasive and seem to be insensitive to what they do to others.

To overcome these "blind spots," people intentionally seek out feedback to learn how others perceive them. This may be part of the reason why people who are self-aware are also better performers.

Self-confidence is the ability to develop a strong sense of one's self-worth and capabilities. People with this competence present themselves with self-assurance; they are decisive and able to make sound decisions despite uncertainties and pressures.

Self-Management

Self-management is about managing one's internal states, impulses and resources—consists of the competencies *emotional self-control, transparency/ trustworthiness, adaptability, achievement orientation, initiative, optimism, and conscientiousness.*

Emotional self-control is the ability to keep disruptive impulses in check. People with this competence manage their impulsive feelings well, they remain positive and composed, even under difficult moments, and they are able to stay focused even under pressure.

At work, the relationship with our boss or supervisor has the greatest impact on our emotional and physical health—giving us often a sense of frustration and helplessness. Quite often a situation can be seen by one person in a negative way—as a destructive threat—but by another in a positive way—as an enlivening challenge. With the right emotional resources, what seems destructive can be taken instead as a challenge, and met with energy and enthusiasm.

Transparency/trustworthiness and conscientiousness is the ability to maintain integrity and to feel responsible for each other. People with this competence act ethically—one of the most important considerations. They succeed to build trust through their reliability, they are open to admit their own mistakes and confront unethical actions in others. They stick to their positions even if they are unpopular. They keep promises and hold themselves accountable if they meet or fail to meet their objectives, and they perform their work in an organized and careful way.

Adaptability is the ability of being open to novel ideas and approaches, and being flexible in responding to change. People with this competence always are open to look for fresh ideas from many different sources and generate new ideas themselves. They often prefer original solutions to problems and are willing to take risks in their thinking. They are flexible to deal with multiple demands and changing priorities, and they are able to quickly adapt their responses, even dropping everything without reservation as realities shift.

Adaptability requires the flexibility to consider various aspects of a given situation. Flexible people have the ability to stay comfortable with uncertainty and remain calm in the face of the unexpected. It is also supported by self-confidence, which helps to quickly adjust their responses. We can see the two extremes. On the one end we have the innovator, who is driven mainly by originality. He is able to quickly simplify seemingly complex problems and adapt to a new situation.

On the other end we find people who don't like to take risks; they may get lost in details and remain critics. Such people, often managers, waste time on evaluating various ideas; they micromanage and put pressure on deadlines, which creates feelings of being suppressed. Such behavior invariably will reduce creativity.

Creativity tends to be enhanced in organizations that have less formality, allow more ambiguous and flexible roles, give workers autonomy, have open flows of information, and operate in mixed or multidisciplinary teams.

Achievement orientation is the ability to continuously strive to improve or meet a standard of excellence. People with this competence are results-oriented; they set themselves ambitious goals, which may include calculated risks.

Example:

ABB, when created, required from business areas to accept only profitable jobs; they also wanted to be world-wide number one in their field.

The calculated risk: entrepreneurial drive demands that people be comfortable taking risks but know how to calculate them carefully.

A passion for feedback: whenever a working group meets regularly to find ways to improve performance, they embody a collective drive to achieve.

The pursuit of information and efficiency: "management by walking around" or encouraging impromptu contacts or informal meetings with people at all levels.

Initiative and optimism is the ability to show proactivity and persistence. People with this competence are ready to seize opportunities. They pursue goals beyond the targets given to them, even if it means cutting red tape or bending rules. They continuously motivate others and are driven more by hope of success than by fear of failure, and they know that although setbacks will always show up, they can be managed.

People with initiative often act early enough to avoid problems before they happen. Bosses who micromanage may seem to have initiative, but they lack a basic awareness of how their actions affect other people.

Social Awareness

Social awareness consists of the competencies *empathy, organizational awareness, and service orientation.* Social awareness is the ability to sense, understand, and react to others' emotions while comprehending social networks.

Empathy is the ability to "read" the emotions of others' feelings, to understand their perspectives, and to take an active interest in their concerns. People with this competence pay attention to emotional signals and listen well. They help out based on understanding other people's needs and feelings.

Organizational awareness is the ability to see what is going on beneath the surface of organizations, to read social and political currents. People with this competence read key power relationships, organizational and external realities, and they understand the forces that shape views and actions of clients, customers, or competitors.

Service orientation is the ability to anticipate, recognize, and meet customers' needs. People with this competence match customers' needs to services and products, they always offer their assistance and are open and capable of knowing what the customer may need next.

Relationship Management

Relationship management refers to how we handle the emotions of others. Ways that we manage relationships include skills how to influence others, developing others through feedback and guidance, being a catalyst for change, resolving conflicts in a mutually agreeable way, and promoting positive relationships.

Leadership is the ability to inspire and guide individuals and groups. People with this competence articulate and arouse enthusiasm for a shared vision and mission—speaking to people's hearts. They can step forward to lead when there is a need, regardless of their position. They guide the performance of others while holding them accountable. As leaders they want to demonstrate an example. The ability to convey emotion convincingly, from the heart, requires that a leader be sincere about the message being delivered.

Influence is the ability to apply effective tactics for persuading others. People with this competence will fine-tune presentations to reach out to the listener. They may use indirect influence to build consensus and support. Sometimes they even orchestrate dramatic events to effectively make a point.

If this competence is still weak, it may show up in various ways: a failure to reach consensus; re-using the one familiar strategy instead of choosing the best for the moment; a stubborn promotion of one point, regardless of differing feedbacks; being more politically than performance oriented managers—they are effective upwards, but poor downwards, and often being self-centered they think more of themselves rather than the organization, having a negative impact.

Developing others is the ability to recognize early others' development needs and to strengthen their abilities. People with this competence show a genuine personal interest in those they guide, and have empathy for and an understanding of their employees. They offer useful feedback and recognize people's needs for further development. One way to promote positive expectations is to let others take the lead in setting their own goals, rather than dictating the terms and manner of their development.

Change catalyst is the ability to initiate or manage change. People with this competence recognize the need for change by challenging the status quo. They champion the change and involve others as well. They can model the change expected of others. In addition to high levels of self-confidence, effective change leaders have high levels of influence, commitment, motivation, initiative, and optimism, as well as an instinct for organizational politics.

Conflict management encompasses the ability to negotiate and resolve disagreements. People with this competence look at a conflict from all sides. They handle difficult people and tense situations tactfully; they bring all disagreements into the open, helping thereby to de-escalate; and they encourage an open discussion with the goal to reach a win-win solution.

Building bonds refers to the ability to promote effective relationships. People with this competence maintain extensive informal networks. They build up an information network, keeping others in the loop, and they generate personal friendships among work associates.

Examples from my own experiences:

The Arlberg-Colloquium in Austria was an annual meeting between scientists and teachers from an Austrian university and the Max Planck Institute: skiing during the day was followed by intense, lively discussions until late at night. Similar events have been common in companies such as ABB and other ones. A good example was presentation courses which each scientist from ABB's Corporate Research Centers was obliged to take.

Teamwork and collaboration is the ability to work together with others towards shared goals. People with this competence balance a focus on the task with giving attention to relationships. They openly share plans, information, and resources, trying to collaborate. They do this in an open, friendly atmosphere, and they are always in search for opportunities for collaboration.

Communication is the ability to listen openly and send convincing messages. People with this competence foster open communication and stay receptive to bad news as well as good news. They can deal with difficult issues in a straightforward way.

Self-test

A simple example of a self-test, which I have used with small groups of students, is to go through the details of each competence and then ask each student to grade himself or herself on a scale from 1 to 3, 1 referring to a low competence level and 3 to a high one. What do we do with the results? The last thing would be to discuss them in a group, because this information should be treated with respect and confidentiality. One approach is to sit privately with each student and discuss of why each came to this conclusion. A teacher who wants to be helpful should try to have a constructive, positive attitude throughout the discussion. My experience is that this process helps bring the two parties closer together and it might also result in the wish to systematically undertake actions to improve specific skill levels. A similar approach can be taken to get to know the members of a work team, thereby improving the relationship between the team leader and the other members. Figure 10.2, still based on Goleman's [6] earlier 5 characteristics shows a couple of results of such a test. One could quickly see that the student's environment made a difference in the self-assessment.

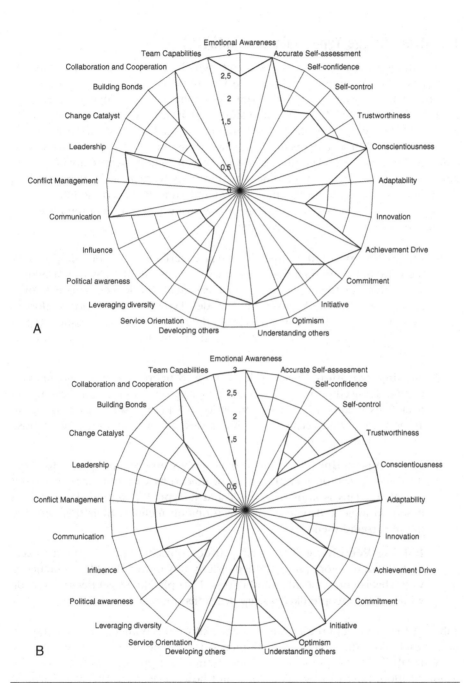

FIGURE 10.2 Self-assessment of emotional intelligence without using a questionnaire: (a) Best in class, outgoing; (b) average performance, withdrawn.

The Myers-Briggs Type Indicator® (MBTI®)

While emotional intelligence is a skill set, which can be improved over time, personalities are preferences with which we are born.

One of the most popular personality assessments is the Myers-Briggs Type Indicator® (MBTI®). C. G. Jung [12] postulated that there are psychological differences and different types among persons. Based on his work, Katharine Cook Briggs and Isabel Briggs Myers designed an assessment tool to identify these psychological differences [13] during World War II. They were convinced that knowledge of personality preferences would help women who were entering the industrial workforce for the first time identify the right wartime job.

Myers-Briggs® assessment uses four interweaving pairs of dichotomics to distinguish between the various preferences:

1. **Extraversion–Introversion**. This pair describes where people prefer to focus their attention and get their energy: from the outer world of people and activity or their inner world of ideas and experiences. This is known as the E and I in the Myers-Briggs® model. Type E people process information on the outside and like to discuss their way through the issues. Type I people prefer to process information on the inside, needing quiet time for thinking.

2. **Sensing–Intuition**. This pair describes how people prefer to take in information: focused on what is real and actual or on patterns and meanings in data. This is known as the S and N. Type S people see what is going on around them while types N are less aware of their environment because they are thinking one thing through.

3. **Thinking–Feeling**. The third pair describes how people prefer to make decisions: based on logical analysis or guided by concern for their impact on others. This is known as the T and F. Type T people make decisions based on logic and type F base decisions on feelings, their own and the people around them.

4. **Judging–Perceiving**. The fourth pair describes how people prefer to deal with the outer world: in a planned orderly way or in a flexible spontaneous way. This is known as the J and P. Type J people like to get decisions made while type P people like to keep decisions open and flexible.

Table 10.1 is a matrix of the 16 MBTI® personality types with a shortened description of each type. Much more detailed information can be found at www.cpp.com.

Your MBTI® profile guides you to understand your preferences how you take in information, how to make decisions, and how to relate to the world around you. The profile informs you why you think and act in certain ways. Having this information, you will be aware how your behaviors affect others, and you can modify these behaviors if the situation so demands. Each of the 16 distinct MBTI

Table 10.1 The 16 MBTI Personality Type Matrix [14]

| | | Sensing Types (S) | | Intuitive Types (N) | |
		Thinking (T)	Feeling (F)	Feeling (F)	Thinking (T)
Introverts (I)	Judging (J)	**ISTJ** Serious, quiet, successful by focus and reliability. Practical, realistic, and loyal. Pursue goals logically and without distractions. Enjoy an orderly work and life.	**ISFJ** Quiet, friendly, and conscientious. Accurate, always work to meet commitments. Considerate and aware of what others think. Looking for harmony at work and at home.	**INFJ** Search to understand ideas, relationships, and motivation of people. Follow clear visions based on firm principles. How best to serve the common good. Decisive in implementation.	**INTJ** Have original minds and are keen to reach their goals. Quickly think in patterns and long-term perspectives. Sceptical, independent, with high standards of competence and performance.
	Perceiving (P)	**ISTP** Quiet, tolerant observers. Act fast to find a workable solution. After careful analysis eager to quickly isolate the core of problems. Organize facts using logical principles.	**ISFP** Quiet, sensitive, kind. Enjoy the moment, need to work within their own space and time-frame. Committed to values and people. Look for agreements and accept others' opinions.	**INFP** Idealistic, lead external lives congruent with values important to them. Curious, flexible. Quickly see new ideas and start to implement them. Interested to understand and help people.	**INTP** Find logic in everything they are interested in. Skeptical, appear critical. Focus more on theories behind ideas than social interaction. Quiet, adaptable, never loose focus.
Extroverts (E)	Perceiving (P)	**ESTP** Flexible, driven mainly to reach results quickly. Not interested in theories, but focus on implementation, on real things. Enjoy activity with others, also material comforts. Learning by doing.	**ESFP** Outgoing and accepting. Enjoy life, people, and material comforts. Work well with others. Use common sense and realism in their work. Flexible, adapt quickly to new environments.	**ENFP** Warmly enthusiastic, always finding new possibilities. Follow patterns with confidence. Looking for and showing appreciation and support. Spontaneous, flexible, can improvise and talk.	**ENTP** Quick, ingenious, and outspoken. Bored by routine. Come up fast in solving new problems, can generate and analyze concepts strategically. Understand well others. Look for new tasks.
	Judging (J)	**ESTJ** Practical, decisive. Quickly implement decisions. Get projects done fast and forcefully, well organized, choosing right people. Will consider details. Follow clear sets of standards, expect others to follow.	**ESFJ** Conscientious, warm-hearted, and cooperative. Establish and work in harmony. Work and help others to finish task accurately and on time. Help where needed, but also expect appreciation.	**ENFJ** Warm, responsive, and responsible. Aware of emotions of others. See and use potential in everyone. Responsive to praise and criticism. Help others in a group, are inspirational leaders.	**ENTJ** Assume leadership fast. Quickly identify mistakes, implement whole systems clarify organization. Use long-term planning. Well informed, expand and pass on information to others. Present ideas clearly.

personality types has its strengths as well as its challenges. The most important point—no one type is better than another.

How should we use the information? First of all, it is important that we can see ourselves from "above"; second, we must realize that there are many types of personalities, and as Leonard and Straus [14] argue, the highest potential for an innovative team comes from a mix of preferences. Knowing that such differences exist puts a burden on the manager to learn to deal with them in a productive process, which they call *abrasion*.

The Hermann Brain Dominance Instrument® (HBDI®)

Earlier research had shown that different parts of the brain have different functions and work independently from each other. Based on 20 years of research on brain dominance, Hermann [11] created a Four-Quadrant Model, from which he derived and tested the Hermann Brain Dominance Instrument® (HBDI®), following a request from GE, where he had the opportunity to test his approach on countless employees.

In his model, Herrmann assumes that each person has different modes of thinking. Depending on which mode is dominating one belongs to one or more of the four groups, each represented by a quadrant. The four different modes are defined by typical ways of thinking and acting:

A. **Upper Left: Analytical thinking**. Key words: logical, factual, critical, technical, and quantitative. Preference to this quadrant means that a person favors activities that involve logical and fact-based information combined with an ability to perceive and express information precisely.

B. **Lower Left: Sequential thinking**. Key words: safekeeping, structured, organized, reliable, detailed, planned. Preference to this quadrant means that a person favors following directions, organized and detail-oriented information and work, step-by-step problem solving.

C. **Lower Right: Interpersonal thinking**. Key words: kinesthetic (involvement of physical activities), emotional, spiritual, sensory, and feeling. Preferred activities in this quadrant are listening to and expressing ideas, looking for personal meaning, emotional input, and interaction with others.

D. **Upper Right: Imaginative thinking**. Key words: visual, holistic, intuitive, innovative, and conceptual. Preference to this quadrant means looking at the big picture, taking initiative, challenging assumptions, applying metaphoric thinking, creative problem solving, and long-term thinking.

The HBDI® study of over two million people around the globe has demonstrated that using a Whole Brain® approach (i.e., an approach that uses all four quadrants), as in the heterogeneous team below, will significantly increase the innovative and creative output of a team. A study conducted by the US Forest Service showed that heterogeneous teams can improve team effectiveness by up to 66% when a team is no larger than seven members and at least one member is a "whole brain thinker." In the HBDI®, each individual has access to all four thinking styles, but no varying degrees.

Figure 10.3 shows how the different preferences combine into four quadrants and how we can use the charts to distinguish among different teams, in this case, between a homogeneous team and a heterogeneous one.

The knowledge of personality types should help you, as a student, teacher, engineer, manager or part of a team, to recognize and appreciate the differences of others. Students with the same preferred thinking style will find it easier to communicate and understand each other, compared to students who may have opposite preferences of thinking.

Which combination of personality types is best suited to achieve the maximum innovation? While it is impossible to come up with a perfect answer, it is important to note that innovation is the result of finding the best solution when different ideas, perceptions, and ways of processing and judging information come together. This, inadvertently, often requires collaboration among people who think differently and might run into a conflict. Disputes become personal, and the creative process breaks down. Managers who dislike disputes might avoid these problems by hiring people with similar personalities. Such a group will easily agree on the right idea but might totally fail to innovate. The manager who decides to bring together different personalities has to figure out the most productive way to create innovation out of the very different ideas.

Many other tools are in use today by human resources departments or offered by consulting services, often the result of the need to differentiate among people. Also, attempts have been made to find statistical correlations between the various techniques. The purpose of this chapter is not to go into deep but perhaps unrewarding analysis of the various approaches but to make you aware that everyone has a unique personality and emotional competencies and you have to find ways to take advantage of these differences.

IMPORTANT ASPECTS IN MANAGEMENT

The Role of Intuition Versus Rational Decision Making

The interesting main conclusion from the case studies about NDC and Metoxit, and most of the other case studies as well, is that rarely could we ever extract a rule-based approach from management literature, but must follow our gut, intuition, applied common sense, and the like. Does this mean to let chaos reign as opposed to a structured, well-organized approach? Reality shows that structure always lies behind each approach, although it differs from case to case. One possible explanation is the variety of the cases: Different approaches are required

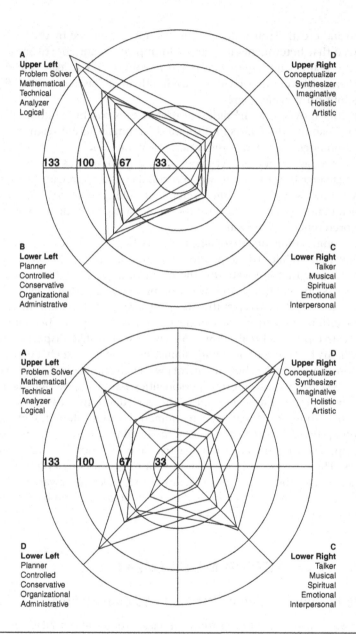

FIGURE 10.3 Examples of a HBDI® Team Profiles homogeneous (top) and a heterogeneous (bottom) team. (The HBDI® and Whole Brain® Model are trademarks of Herrmann International, Inc. and are reproduced with written permission for display in this text. Copyright © 2009 Herrmann International, Inc. All Rights Reserved.)

when you develop a new processing technique for an established product, a new product from a new material, or a promising new material appears and you are look for an application. The complexity shows up when you assemble a team of people with different skill sets and conduct a brainstorming session. In no time, the team will come up with hundreds of ideas. The art is now to assign values to each idea and choose the most likely winner. The fact that a majority of the team selects one idea is still no guarantee that it will successfully become a product or a new business. In other words, each decision faces a probability of success but also a probability of failure. As there are many decisions to take, you have to multiply the probabilities and come up with a result from practical experience:

> Only 1 out of 1000 ideas will succeed.
> Only 5% of new businesses will make it to the market.
> The good news is that making mistakes is an important part of the learning process.

Undoubtedly, lots of tools around will help you in making the right decision, but they do not eliminate the risks. Let me explain this to you with a simple example. Someone developed a new manufacturing process that allows producing complex structural components of net shape and high quality. The process has been well established, and lots of companies are competing as suppliers of the manufactured parts. Some customers ask for larger components that require a much larger (and much more expensive) machine, which no one has used before. Shall we take the risk and purchase a very expensive new machine? Will we recover the high investment costs? Will we have a guaranteed customer base? Will we be able to find new customers? You have to make assumptions about the right answer to the last questions, because they lie in the future.

This is the right place to discuss the role of intuition [15], and emotion, in making management decisions. Over the past 30–40 years, we have seen the evolution of a wide range of tools to aid the logical decision-making process, the tools of operations research and management science, but above all, the technology of expert systems. These tools have been applied mostly to decisions that are well structured and quantitative. The logical decision-making process relies mostly on logic and quantitative analysis. You establish all possible options and carefully analyze them. You formulate the main criteria for judging the expected outcomes of your options and assign certain weights to each of the criteria to reflect their relative importance. Based on the expected outcomes and their weights, you can rate the various options and finally select the one with the highest rating. Rational analysis can play an important role in many situations, especially those where you have clear criteria and have to deal with extensive quantitative data, both technical and financial. Often, this leads to the conclusion that only those who master these tools can make the right decisions, and lots of consulting companies use these tools as their major knowledge base. But these tools have had less impact on decision making that is loosely structured, intuitive, and qualitative. And they had even less impact on decisions resulting from face-to-face interactions between a manager and coworkers.

In still more business situations, the rational decision making becomes impractical. In nonrational or intuitive decision-making processes, response times in decision making are very short, too short to allow for a systematic quantitative analysis of all potential variants. Still, experienced decision makers have the right experience and confidence to make such rapid decisions. Explanations are based on physiological factors: the 'two-brains' hypothesis [10], which argues that rational and intuitive processes are so different that they are carried out in different parts of the brain. We saw this earlier, when describing the Herrmann Brain Dominance Instrument® (HBDI®).

It is interesting to note that studies have been made on the different speeds with which experienced and less experienced chess players conduct the game [16]. The so-called grand masters are not only much faster playing their own game, they can play simultaneously against up to 50 opponents. Tests have shown that this is based on their memory of up to 50,000 patterns on a chess set, to which they can respond instantaneously. Amateurs with much less experience have to go through each pattern in a learning process and therefore use much more time and "logical thinking."

The same concept applies to the use of intuition in management. Like the chess player, a manager accumulates a large set of scenarios and responses to business questions. This makes it possible to diagnose problems and come up with solutions rapidly and intuitively without being able to specify how he or she attained the result. It is the purpose of expert systems to store these "patterns" of information and facilitate and accelerate the decision-making process. The simplest way to understand how intuition works is to think of it as an advanced pattern recognition device. Your subconscious mind somehow finds links between your new situation and various patterns of your past experiences. Although you do not remember enough of the details to go through the process of analytical reasoning, your subconscious mind remembers the patterns learned. It can rapidly project your new situation onto those patterns and send you a message, more likely to be expressed in the language of your feelings.

What managers actually do is often different from what they should do either rationally by analysis or intuitively. Often uncertainty leads them to postpone a decision. What should we have done?

Perhaps the biggest influence a manager can have on the problem-solving style of an organization is to make the best response to problems. These are some of the key principles to be followed:

- Solving the problem takes priority over looking backward to the causes and searching for the scapegoats.
- Accepting personal responsibility for finding and proposing solutions instead of delegating this task to someone else, often up in the organization.
- Accepting personal responsibility for implementing action solutions.

An effective manager commands a whole range of management skills, from analytical to intuitive and applies them as they become appropriate.

Here are some examples of patterns:

- Understanding that materials are part of a system and learning about the system (for example, to make the right decisions in developing high-temperature materials for gas turbines, you have to understand the basics of a gas turbine, the "system").
- Knowledge in interdisciplinary fields (making the right decision in nano-technology is easier for people with a background in physics, materials science, and life sciences).
- Building the right team to cross the various skill levels.

Once you have gone through the learning process of understanding these patterns, it will take virtually no time to make the right decisions in similarly structured cases.

Compare innovation with navigating a kayak on a river or through tidal currents [17]. Very often, a small river may start up in the mountains and rush down rapidly, later reaching a quieter and wider bed to flow. Continuous innovation in larger, established companies often resembles navigating a boat on a large river, flowing smoothly, a very safe, predictable course. Things are very different in the earlier sections, where water rushes down rocks or, due to a change in tides, unexpected cross-currents turn up and great skills in navigation are required. Still, the probability that you might tip over is quite substantial. If you have considered such accidents to happen, then you may have learned before how to either circumnavigate unforeseen obstacles or rescue yourself once you tipped over. Still, tipping over is a valuable experience, from which you learn that you can deal with setbacks coming from unforeseeable events.

Here are just a few of the unknown obstacles which you have to navigate around:

- Are you starting out with the *right new technology*? Or, will you have to modify your approach by incrementally improving the existing technology? Is the discovery of new high-temperature superconductors with, until now, unknown properties a guarantee for success in the market?
- Should you participate in *government-funded projects*, which save you some money but may slow you down and force you to take directions you cannot change freely, or should you get *internal funding*, either from corporate research (in a large corporation) or directly from the interested business unit?
- Which *organizational approach* should you use? A step-by-step approach, like climbing a ladder or a more adventurous one, allowing you to take shortcuts.
- Do you rigidly follow a tool-based *logic approach* or do you allow *intuition* to play a major role?

- Since *speed* is always important, how can you make sure that you achieve your targets as fast as possible?
- Is the *original market* you chose as driver for your project the right one, does it have to be replaced by another one, or do you need additional markets to secure success?
- Do you have the *right personal skill sets*? Is it better to have the key people within your organization or can you see advantages in buying parts of the technology and other skills from outside to reach your target faster and at lower cost?
- Everything can, and in most cases will, change during the course of developing new product or businesses, and the biggest danger to be aware of is to stick rigidly to the original plan. While this requires flexibility to adjust your strategy, you must not lose sight of the main components of your vision.

Managing People Well

A good manager often has a clear vision of what the future will be. While there is never a guarantee that this vision will become true, it helps provide a focus for the team's activity, which leads to sustained long-term motivation and a united team. A vision has to be something sufficiently interesting to provide the team with a sense of common purpose. Two things are important in this process:

- When you decide on a vision, it should be the direction in which your team should move.
- You have to communicate that vision to your team.

In my first position as a lab manager of the former BBC central laboratory, the vision was based on the three-layer model of R&D (basic research at the corporate research center, applied research at the central laboratory, and product development in the development departments of the business units). The vision was to transform the central lab from its old role as a service lab to one doing applied research, a very difficult task that would require hiring lots of new staff and transferring some activities to the business units. I called the process of communicating this vision to the employees *surprise-free management*—everyone should understand and support the need for change. Communication was one thing, making it work was another challenge, since the established staff did not support the vision, neither did the manager I was reporting to at the time. I never lost track of my vision, but it took three years to create a new team and make it work enthusiastically. During that time, lots of mistakes were made, which prompted two internal consultants who had helped to organize a large workshop to advise me to give up my vision. I refused, which was the right move. A few years later, I got a new boss who had heard about my experience as a lab manager, and I was sent to the United States for two years to set up a new laboratory, partially by combining several dispersed development units. Now, I had learned from my previous mistakes, having acted authoritarian and too slowly. My new approach was this:

- Formulate a vision and communicate it clearly, stressing that it could be made to work only when we all worked together.
- Try to do it as fast as possible.

Four teams were created, one for each future department, and each was given a maximum amount of freedom. After a couple of weeks, I asked for their opinion of how each department should be organized and major issues to address during work. This undoubtedly led to a huge amount of motivation, and after three weeks, the new lab organization and its main research topics were established, boggling the mind of my supervisor.

The Importance of Delegation

We can look at delegating work and responsibilities from two angles: On the negative side, the manager wants to make life easier for by unloading work and putting it on the shoulders of others. Of course he can and will blame them if they fail. On the positive and more important side, a manager empowers and motivates the team to feel responsible for its own work and take more initiative to realize their full potential.

Delegation is primarily about entrusting your authority to others. Another justifier for delegation is the reverse view: Many of us do not want to be controlled or micromanaged by our boss. Some of us intuitively know that we can perform best when left on our own as much as possible, assuming we are given the time to gain experience in making our own decisions.

To enable others to do a job for you, you must ensure that

- They clearly understand what you want.
- They have the competence to do it.
- You have the competence to judge beyond your own experience level.
- They feel comfortable in achieving the objective.
- You do not lose control.

From my experience, I can list three examples:

- Martin, a brilliant engineer, who was reporting to me, one day, told me that he was going to leave his job, because he felt too micromanaged by me, indicating that he knew many things better. At that time, we started to think about getting a low-cost press for research on isothermal forging. I asked him what he would like to do, if he were free to decide; and he came up with a completely novel idea how to build the lowest-cost press, which we did successfully.
- Erwin, my first technician, turned out not only an extremely knowledgeable person but also someone interesting to discuss all aspects of life with. I was still intuitively convinced that I had to know the solution for each task before I could ask him to work on it. One day he said, "could you please just tell me where you want me to get, but don't tell me how to get there."

By following his request, I saw that I had empowered him to do the most difficult assignments for me.

- I soon became convinced that the best innovation comes from working in interdisciplinary fields and putting together a new team of specialists with the right skill sets. Examples of successful world-class research were isothermal forging of Ti alloys and development of new high-temperature shape-memory alloys. In both cases, a person or a small team of people with the right skill set were selected and given a maximum of freedom about how they wanted to execute the project. *Success* could be defined as the moment, when the originator of the idea, the manager, no longer was needed to bring the project to completion. In the world of research, where your accomplishments are proven to the outside world by publications and presentations at conferences, it is not an easy decision to also delegate the deserved fame to those who eventually were successful in completing the task.

Of course, you always want to see a positive outcome from the task you delegated. If good results are achieved, then it helps to keep your staff motivated by giving them all the credit they deserve. Statistically, however, failures occur more often and have to be dealt with. This is the moment when you remember your responsibility, since it was you who delegated the task in the first place. It is most important to sit together in a relaxed mode and try to find the reason for the failure by better understanding the problem, because only then will it be possible to resume the work and bring it to a successful conclusion.

The Contribution of Consultants

The previous discussion on the value of using a good mix of intuition and rational thinking in the decision-making process can be applied to consultants as well. The task of a consultant is to offer specialized services to a customer who either does not have the experience or does not want to build it up or to contribute as staff in time-limited job assignments. The best consultant obviously is the one who brings both long-term experience and the aforementioned mix in making decisions. But, in numerous cases and my experience, consultants focused too much on logical decision-making processes or had become specialists in using certain tools. Any deviation from the knowledge behind the tools used led to erring around in unknown territory. Many consultants developed special ways to express themselves, often using complex vocabulary to convey the impression of being in possession of highest learning.

As more and more information is generated about materials, processing and cost analysis, existing and potential new markets, it is obvious that more and more of the available knowledge is interwoven into tools available to consultants. An example is the recently developed investment methodology for new materials (IMM) by Maine and Ashby [18]. The authors argue rightly that the time from inventing a new material to its commercial application is often 15 years. In some of the case studies discussed in this book, even 20 years is quite common. It must be understood that innovations in new materials applications have been considered a high-risk investment. Therefore, quite often, R&D is left to big companies that have the resources, and government funding is often sought after to reduce the risk. Reality is, however, that many innovations originate in small and medium enterprises, because a few individuals are strongly driven by a vision to make the innovation a success. IMM brings all known methodologies together, dividing them into three segments

- **Viability of the product**. This involves a balance between modeling the technical performance, the economics of production, and the market value.
- **Market assessment**. This consists of techniques for identifying promising market applications and forecasting future volume of production.
- **Value capture**. The likelihood that the innovation leads to value capture is assessed through an analysis of industry structure, organizational structure, intellectual property issues, and the planned market approach.

To sum it up in one sentence, if a product can be made economically and brings value to the market, if there is a substantial market, and if the products, protected by intellectual property rights, generate good profits, then the innovation will be a success.

Let us compare a consultant using such a tool with a chess player. A consultant who has gained pattern thinking through many years of experience will help the client speed up the process and reduce risks and uncertainty. If the consultant is a novice in the field, then the opposite will happen: too many questions asked, too many "bureaucratic" steps required from the client. Tools are important, because they contain lots of accumulated knowledge, but they should always remain what the word indicates—a tool to navigate through a chaotic environment.

Consultants are best, when they are involved in human resource assignments, such as teaching teamwork, presentation skills, project management, or quality control and all those functions that are mainly tool based. Much more is demanded from a consultant when experience and intuition are a must in the decision-making process. The other issue with working with consultants is leadership. Is the consultant the leader or will you remain the leader, even if you lack specialized knowledge? In my experience, the second option is the more realistic one, because someone within a company has to maintain the leadership role throughout the whole process.

SUMMARY

Many similar needs are shared by mountain climbing and leadership in management:

- A willingness to take on risk in an uncertain environment.
- Building trust in the climber ahead of you and responsibility for the person behind you.

- Effective communication.
- Not everyone wants or can be an effective leader in a business environment. *People are different from each other.*

Psychometric tools have been developed to measure our capabilities and learn about our preferences. Three of them are discussed to some extent:

- **Emotional Intelligence (EI).** This is more important than IQ to a successful manager. A simple self-test is described.
- **Myers-Briggs Type Indicator® (MBTI®).** Based on Jung's earlier work, 16 personality types are identified. There are no good or bad ones, all have the same potential. Although it is impossible to come up with an exact description of one's personality, the test allows one to reflect about the complexity of dealing with different types and learn how to manage likely friction between very different types.
- **Hermann Brain Dominance Instrument® (HBDI®).** There are four ways of thinking: analytical, sequential, interpersonal, and imaginative. Mixing different types in a team leads to more innovation than a team consisting of similar types and provides an easy way for a team to review a decision or an idea by "visiting" each of the four quadrants.

One of the main findings from the case studies, especially Metoxit, is the importance of intuition versus rational decision making. To be a successful manager, one has to assemble the right team of good people and learn how to delegate and empower.

The role of consultants is described briefly. Mostly, they possess and apply tools, which can be helpful. One of them, the investment methodology for materials, is highlighted as an example. If used by an experienced consultant, who has acquired the skills of pattern thinking, the contribution can be very valuable; less-experienced ones may create a lengthy, complex process without adding value.

REFERENCES

[1] J. Sims, available at http://undergrad.wharton.upenn.edu/experience.

[2] D. Mazur, available at www.nationalgeographic.com/adventure/everest/dan-mazur-lincoln-hall-video.html.

[3] Available at www.tms-international.com/News/Pharma_marketing_-_psychometric_tools.pdf.

[4] W.L. Payne, A study of emotion: Developing emotional intelligence; self integration; relating to fear, pain and desire, Dissertation Abstracts International 47 (1983–1986) 203A. (University microfilms No. AAC 8605928).

[5] P. Salovey, J.D. Mayer, "Emotional intelligence," Imagination, Cognition, and Personality 9 (1990) 185–211.

[6] D. Goleman, Emotional intelligence, Bantam Books, New York, 1995.

[7] D. Goleman, Working with emotional intelligence, Bantam Books, New York, 1998.

[8] D. Goleman, Emotional intelligence, Bantam Books, New York, 2006.

[9] Van der Zee, M. Thiis, L. Schakel, The relationship of emotional intelligence with academic intelligence and the Big Five, European Journal of Personality 16 (2) (2002) 103–125.

[10] I. Briggs Myers, Introduction to Type, sixth ed., Consulting Psychologists Press, Inc., Palo Alto, CA, 1993.

[11] N. Herrmann, The Whole Brain Business Book, McGraw-Hill, New York, 1996.

[12] C.G. Jung, Psychological Types, in: Collected Works of C. G. Jung, vol. 6, Princeton University Press, Princeton, NJ, 1971.

[13] I. Briggs Myers, Gifts Differing: Understanding Personality Type, Davies-Black Publishing, Mountain View, CA, 1995.

[14] D. Leonard, S. Straus, Putting Your Company's Whole Brain to Work, Harvard Business Review on Knowledge Management (1998).

[15] H.A. Simon, Making Management Decisions: The Role of Intuition and Emotion, Academy of Management Executive (1987) 57–64.

[16] A.J. Slywotzky, D.J. Morrison, Pattern thinking: A strategic shortcut, Strategy and Leadership 28 (1) (2000) 12–17.

[17] P.B. Vaill, Managing as a Performing Art, Jossey-Bass Publishers, San Francisco, 1991.

[18] E.M.A. Maine, M.F. Ashby, An investment methodology for materials, Mater. Des. 23 (2002) 297–306.

Index

Printed in the United States
By Bookmasters